Geometry

LARSON
BOSWELL
STIFF

Applying • Reasoning • Measuring

Chapter 11 Resource Book

The Resource Book contains the wide variety of blackline masters available for Chapter 11. The blacklines are organized by lesson. Included are support materials for the teacher as well as practice, activities, applications, and assessment resources.

McDougal Littell
A HOUGHTON MIFFLIN COMPANY
Evanston, Illinois • Boston • Dallas

Contributing Authors

The authors wish to thank the following individuals for their contributions to the Chapter 11 Resource Book.

Eric J. Amendola
Patrick M. Kelly
Edward H. Kuhar
Lynn Lafferty
Dr. Frank Marzano
Wayne Nirode
Dr. Charles Redmond
Paul Ruland

ISBN: 0-618-02074-8

789-VEI- 04

Contents

11 *Area of Polygons and Circles*

Chapter Support	1–8
11.1 Angle Measures in Polygons	9–22
11.2 Areas of Regular Polygons	23–37
11.3 Perimeters and Areas of Similar Figures	38–52
11.4 Circumference and Arc Length	53–66
11.5 Areas of Circles and Sectors	67–82
11.6 Geometric Probability	83–97
Review and Assess	98–111
Resource Book Answers	A1–A9

Contents

Chapter Support Materials			
Tips for New Teachers	p. 1	Prerequisite Skills Review	p. 5
Parent Guide for Student Success	p. 3	Strategies for Reading Mathematics	p. 7

Lesson Materials	11.1	11.2	11.3	11.4	11.5	11.6
Lesson Plans (Reg. & Block)	p. 9	p. 23	p. 38	p. 53	p. 67	p. 83
Warm-Ups & Daily Quiz	p. 11	p. 25	p. 40	p. 55	p. 69	p. 85
Activity Support Masters			p. 41			
Alternative Lesson Openers	p. 12	p. 26	p. 42	p. 56	p. 70	p. 86
Tech. Activities & Keystrokes	p. 13	p. 27		p. 57	p. 71	p. 87
Practice Level A	p. 14	p. 30	p. 43	p. 59	p. 75	p. 89
Practice Level B	p. 15	p. 31	p. 44	p. 60	p. 76	p. 90
Practice Level C	p. 16	p. 32	p. 45	p. 61	p. 77	p. 91
Reteaching and Practice	p. 17	p. 33	p. 46	p. 62	p. 78	p. 92
Catch-Up for Absent Students	p. 19	p. 35	p. 48	p. 64	p. 80	p. 94
Coop. Learning Activities	p. 20					
Interdisciplinary Applications	p. 21		p. 49		p. 81	
Real-Life Applications		p. 36		p. 65		p. 95
Math and History Applications			p. 50			
Challenge: Skills and Appl.	p. 22	p. 37	p. 51	p. 66	p. 82	p. 96

Review and Assessment Materials			
Quizzes	p. 52, p. 97	Alternative Assessment & Math Journal	p. 106
Chapter Review Games and Activities	p. 98	Project with Rubric	p. 108
Chapter Test (3 Levels)	p. 99	Cumulative Review	p. 110
SAT/ACT Chapter Test	p. 105	Resource Book Answers	p. A1

Contents

Descriptions of Resources

This Chapter Resource Book is organized by lessons within the chapter in order to make your planning easier. The following materials are provided:

Tips for New Teachers These teaching notes provide both new and experienced teachers with useful teaching tips for each lesson, including tips about common errors and inclusion.

Parent Guide for Student Success This guide helps parents contribute to student success by providing an overview of the chapter along with questions and activities for parents and students to work on together.

Prerequisite Skills Review Worked-out examples are provided to review the prerequisite skills highlighted on the Study Guide page at the beginning of the chapter. Additional practice is included with each worked-out example.

Strategies for Reading Mathematics The first page teaches reading strategies to be applied to the current chapter and to later chapters. The second page is a visual glossary of key vocabulary.

Lesson Plans and Lesson Plans for Block Scheduling This planning template helps teachers select the materials they will use to teach each lesson from among the variety of materials available for the lesson. The block-scheduling version provides additional information about pacing.

Warm-Up Exercises and Daily Homework Quiz The warm-ups cover prerequisite skills that help prepare students for a given lesson. The quiz assesses students on the content of the previous lesson. (Transparencies also available)

Activity Support Masters These blackline masters make it easier for students to record their work on selected activities in the Student Edition.

Alternative Lesson Openers An engaging alternative for starting each lesson is provided from among these four types: ***Application, Activity, Geometry Software,*** or ***Visual Approach.*** (Color transparencies also available)

Technology Activities with Keystrokes Keystrokes for Geometry software and calculators are provided for each Technology Activity in the Student Edition, along with alternative Technology Activities to begin selected lessons.

Practice A, B, and C These exercises offer additional practice for the material in each lesson, including application problems. There are three levels of practice for each lesson: A (basic), B (average), and C (advanced).

Contents

Reteaching with Practice These two pages provide additional instruction, worked-out examples, and practice exercises covering the key concepts and vocabulary in each lesson.

Quick Catch-Up for Absent Students This handy form makes it easy for teachers to let students who have been absent know what to do for homework and which activities or examples were covered in class.

Cooperative Learning Activities These enrichment activities apply the math taught in the lesson in an interesting way that lends itself to group work.

Interdisciplinary Applications/Real-Life Applications Students apply the mathematics covered in each lesson to solve an interesting interdisciplinary or real-life problem.

Math and History Applications This worksheet expands upon the Math and History feature in the Student Edition.

Challenge: Skills and Applications Teachers can use these exercises to enrich or extend each lesson.

Quizzes The quizzes can be used to assess student progress on two or three lessons.

Chapter Review Games and Activities This worksheet offers fun practice at the end of the chapter and provides an alternative way to review the chapter content in preparation for the Chapter Test.

Chapter Tests A, B, and C These are tests that cover the most important skills taught in the chapter. There are three levels of test: A (basic), B (average), and C (advanced).

SAT/ACT Chapter Test This test also covers the most important skills taught in the chapter, but questions are in multiple-choice and quantitative-comparison format. (See *Alternative Assessment* for multi-step problems.)

Alternative Assessment with Rubrics and Math Journal A journal exercise has students write about the mathematics in the chapter. A multi-step problem has students apply a variety of skills from the chapter and explain their reasoning. Solutions and a 4-point rubric are included.

Project with Rubric The project allows students to delve more deeply into a problem that applies the mathematics of the chapter. Teacher's notes and a 4-point rubric are included.

Cumulative Review These practice pages help students maintain skills from the current chapter and preceding chapters.

Tips for New Teachers

For use with Chapter 11

LESSON 11.1

TEACHING TIP The Activity on page 661 gives students another practical experience of discovering patterns. The results of the relationship they explore are stated as Theorem 11.1 on page 662. Students should recognize that the number of sides and the number of interior angles are the same in a convex polygon. Point out that the formulas given in Theorems 11.1 and 11.2 can be used to find the number of sides in a polygon as well as the angle measures. Example 2 on page 662 illustrates this use.

TEACHING TIP In addition to Example 3 on page 663 you might want students to realize that an exterior angle and its adjacent interior angle form a linear pair. Thus, they are supplementary and the sum of their measures is 180°.

INCLUSION Students with learning difficulties would benefit from a hands-on experience with the diagrams shown at the top of page 663. Have them actually do the steps listed. This will help them understand the meaning of an exterior angle. It will also help them see that only one angle at each vertex is used in Theorem 11.2.

LESSON 11.2

COMMON ERROR Refer students to the Study Tip on page 669. Students must be careful when writing formulas or equations where a variable and a radical sign appear together. Sloppy or careless handwriting may make a variable appear as if it were under a radical symbol when it should not be there. You may want to encourage students to write expressions with the variables before the radical expression. Also caution students about carefully writing the letter *s*, either upper or lower case, because it may look very much like the number five.

TEACHING TIP Explain the process used to solve Example 3 on page 671. Analyze the given information first. Students should understand why it was necessary to find the measure of the central angle. They need to recall properties of isosceles triangles and recall that trigonometric ratios can be used to find unknown lengths. Let students know how you want them to give approximate answers to related problems.

LESSON 11.3

TEACHING TIP When you introduced Lesson 8.3, you may have discussed the ratio of the areas of similar polygons as equaling the square of the ratios of corresponding side lengths. In any case, Activity 11.3 on page 676 supports Theorem 11.5 on page 677.

LESSON 11.4

TEACHING TIP As you explain the Arc Length Corollary on page 683, students should recognize that dividing the degree measure of the arc by 360° represents the fractional part of the circle for that arc. It is that fraction when multiplied by the circumference that gives the linear arc length.

TEACHING TIP The Study Tip on page 684 gives directions for using π and rounding with a calculator. This may not agree with what you suggested to students earlier in this chapter. Confirm your expectations with the students. Be aware that the textbook uses a variety of approaches.

COMMON ERROR Example 4 on page 685 illustrates a problem where attention must be paid to the units of measure that are being used. Students frequently ignore measurement units or fail to recognize that they are different within a particular problem. Stress that students should write their solutions to include units of measure, as in this example, to avoid such errors.

LESSON 11.5

TEACHING TIP Students have had experience finding measures for radii, diameters, areas, and circumferences, but this may be the first time that they have worked with a sector of a circle. They can think of a sector of a circle as a piece of pie.

TEACHING TIP Finding areas of regions or shaded regions requires analyzing the problem and any given diagrams. As in Examples 4 and 5 on page 693, have students write an equation for the

Tips for New Teachers

For use with Chapter 11

areas using words and symbols before writing algebraic equations. For problems similar to Example 6 on page 694, point out that it might also be necessary to draw additional diagrams showing the various portions of the shaded shape.

Lesson 11.6

COMMON ERROR Students may tend to base their answers to problems like Example 1 on page 699 on the number of points showing on the line segments rather than the lengths of the line segments. Stress the wording of both probability equations given above Example 1.

TEACHING TIP Remind students that a probability can be written as a fraction, a decimal, or a percent. This is illustrated by the statement in the solutions to Examples 2, 3, and 4, on pages 700 and 701. Let students know what you want to see as a final answer to problems similar to these.

Outside Resources

BOOKS/PERIODICALS

Rulf, Benjamin. "A Geometric Puzzle That Leads to Fibonacci Sequences." *Mathematics Teacher* (January 1998); pp. 21–23.

ACTIVITIES/MANIPULATIVES

Naraine, Bishnu and Emam Hoosain. "Investigating Polygonal Areas: Making Conjectures and Proving Theorems." *Mathematics Teacher* (February 1998); pp. 135–142, 148–150.

SOFTWARE

Geometry Through the Circle with The Geometer's Sketchpad. Blackline masters and Macintosh and Windows disks with sample sketches and scripts. Berkeley, CA. Key Curriculum Press.

VIDEOS

Basic Geometry Video. Finding areas, perimeters, volumes and coordinate points. Vernon Hills, IL; ETA.

NAME ______________________________ DATE __________

Parent Guide for Student Success

For use with Chapter 11

Chapter Overview One way that you can help your student succeed in Chapter 11 is by discussing the lesson goals in the chart below. When a lesson is completed, ask your student to interpret the lesson goals for you and to explain how the mathematics of the lesson relates to one of the key applications listed in the chart.

Lesson Title	*Lesson Goals*	*Key Applications*
11.1: Angle Measures in Polygons	Find the measures of interior and exterior angles of polygons. Use measures of angles of polygons to solve real-life problems.	• Softball Home Plate • Stained Glass Windows • Houses and Tents
11.2: Areas of Regular Polygons	Find the area of an equilateral triangle. Find the area of a regular polygon.	• Foucault Pendulums • Basaltic Columns • Telescopes
11.3: Perimeters and Areas of Similar Figures	Compare perimeters and areas of similar figures. Use perimeters and areas of similar figures to solve real-life problems.	• The Chicago Board of Trade • Taliesin West Triangular Pool • Fort Jefferson
11.4: Circumference and Arc Length	Find the circumference of a circle and the length of a circular arc. Use circumference and arc length to solve real-life problems.	• Tire Revolutions • Track Length • Bicycles
11.5: Areas of Circles and Sectors	Find the area of a circle and of a sector of a circle. Use areas of circles and sectors to solve real-life problems.	• Boomerangs • Lighthouses • Viking Longships
11.6: Geometric Probability	Find a geometric probability. Use geometric probability to solve real-life problems.	• Trolley Ride • Ship Salvage • Archery

Test-Taking Strategy

Staying Relaxed is the test-taking strategy featured in Chapter 11 (see page 712). Have your student practice techniques to stay relaxed and control panic during a test. Encourage your student to put his or her pencil down and take a few deep breaths if he or she starts to panic. Remind your student that no single test will determine his or her future.

NAME ______________________________ DATE __________

Parent Guide for Student Success

For use with Chapter 11

Key Ideas Your student can demonstrate understanding of key concepts by working through the following exercises with you.

Lesson	Exercise
11.1	The measure of each interior angle of a regular polygon is 108°. What type of regular polygon is it?
11.2	A floor tile is in the shape of a regular hexagon that is 5 inches on a side and has a radius of 5 inches. What is the approximate area of the tile? Find the minimum number of tiles it would take to cover the floor of a 5-foot by 8-foot entry.
11.3	The Square Pizza Place sells pizzas that are square. The regular pizza is 11 inches on a side and costs $8.99. At the same rate, how much should the Square Pizza Place charge for the Jumbo, which is 16.5 inches on a side?
11.4	An archeologist finds a part of a circular pottery plate. The outer rim of the piece is about 39 centimeters in length. The archeologist determines the outer rim represents an 80° arc. What was the original circumference of the plate?
11.5	A goat is tethered by a rope to the corner of a building and can graze in an area that is a circle minus a 90° sector cut by the building. If the rope is 12 feet long, what is the approximate area where the goat can graze?
11.6	A dart is equally likely to land on any point on a 15-inch by 20-inch dart board. Find the probability it lands in the bull's eye, which is a circle of diameter 6 inches in the center of the board.

Home Involvement Activity

You Will Need: Drawing materials
Directions: Work together to design a dream kitchen with a semicircular, bay breakfast nook. Find the floor area of the kitchen. Add cabinets, appliances, and possibly a cooking island to your design plans. Find the area that would need flooring. Research the cost of the flooring type you would choose if money was not a factor. Then calculate the cost for flooring in your dream kitchen.

Answers

11.1: pentagon **11.2:** about 65 in.^{2}; 89 tiles **11.3:** about $20.23 **11.4:** about 176 cm
11.5: about 339 ft^2 **11.6** about 9.4%

NAME ______________________________ DATE ____________

Prerequisite Skills Review

For use before Chapter 11

EXAMPLE 1 Finding an Angle Measure

Find the measure of $\angle B$ and the exterior angle at each vertex.

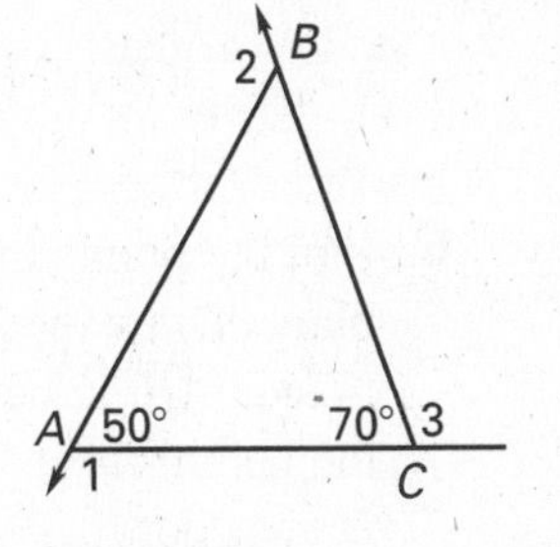

SOLUTION

$m\angle A + m\angle B + m\angle C = 180°$	Triangle Sum Theorem
$50° + m\angle B + 70° = 180°$	Substitute.
$m\angle B + 120° = 180°$	Simplify.
$m\angle B = 60°$	Simplify.
$m\angle 1 + m\angle A = 180°$	Linear Pair Postulate
$m\angle 1 + 50° = 180°$	Substitute.
$m\angle 1 = 130°$	Simplify.
$m\angle 2 = m\angle A + m\angle C$	Exterior Angle Theorem
$m\angle 2 = 50° + 70°$	Substitute.
$m\angle 2 = 120°$	Simplify.
$m\angle 3 + m\angle C = 180°$	Linear Pair Postulate
$m\angle 3 + 70° = 180°$	Substitute.
$m\angle 3 = 110°$	Simplify.

Exercises for Example 1

Find the measure of $\angle B$ and the exterior angle at each vertex.

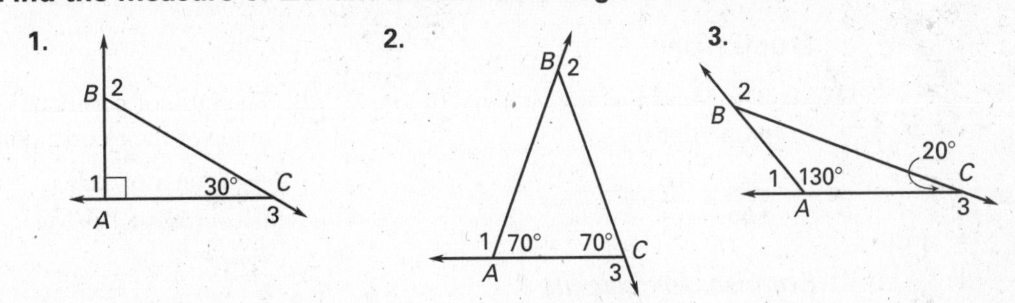

EXAMPLE 2 Area of a Triangle

a. Find the area of a triangle with height 3 in. and base 8 in.

b. Find the area of the triangle defined by $A(2, 2)$, $B(11, 2)$, and $C(5, 6)$.

NAME ______________________ DATE ________

Prerequisite Skills Review

For use before Chapter 11

SOLUTION

a. $A = \frac{1}{2}(\text{base})(\text{height})$ Write formula.

$= \frac{1}{2}(8)(3)$ Substitute.

$= 12 \text{ in.}^2$ Simplify.

b. Plot the points in a coordinate plane. Draw the height from C to side $\overline{AB}$. Label the point where the height meets $\overline{AB}$ as D. Point D has coordinates (5, 2).

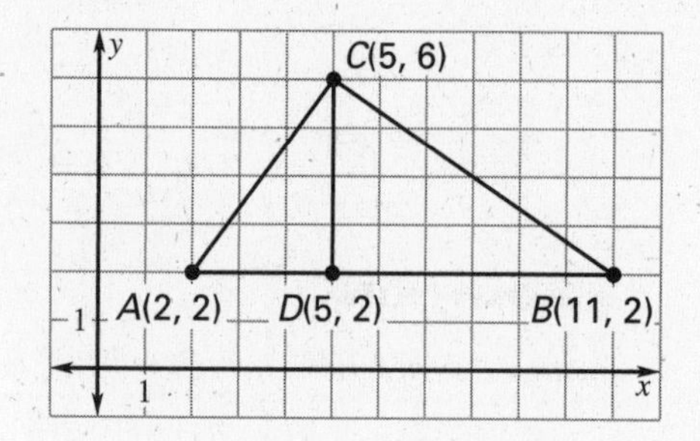

base: $AB = 11 - 2 = 9$

height: $CD = 6 - 2 = 4$

$A = \frac{1}{2}(\text{base})(\text{height})$

$= \frac{1}{2}(9)(4)$

$= 18$ square units

Exercises for Example 2

Find the area of the triangle described.

4. Triangle with height 7 cm and base 4 cm

5. Triangle with height 5 in. and base 9 in.

Draw the triangle in a coordinate plane and find its area.

6. Triangle defined by $A(-1, 3)$, $B(6, 3)$, and $C(3, 10)$

EXAMPLE 3 *Ratios of Similar Triangles*

If $\triangle JKL \sim \triangle MNO$, $JL = 10$, $MO = 15$, find each ratio.

a. $\dfrac{JL}{MO}$ **b.** $\dfrac{\text{perimeter of } \triangle JKL}{\text{perimeter of } \triangle MNO}$

SOLUTION

a. The corresponding side lengths are proportional.

$\dfrac{JL}{MO} = \dfrac{10}{15} = \dfrac{2}{3}$

b. The ratio of perimeters is equal to the ratios of their corresponding side lengths.

$\dfrac{\text{perimeter of } \triangle JKL}{\text{perimeter of } \triangle MNO} = \dfrac{2}{3}$

Exercises for Example 3

If $\triangle RST \sim \triangle XYZ$, $ST = 12$, $YZ = 20$, find each ratio.

7. $\dfrac{ST}{YZ}$ **8.** $\dfrac{YZ}{ST}$ **9.** $\dfrac{RT}{XZ}$

10. $\dfrac{\text{perimeter of } \triangle XYZ}{\text{perimeter of } \triangle RST}$ **11.** $\dfrac{YX}{SR}$ **12.** $\dfrac{XZ}{RT}$

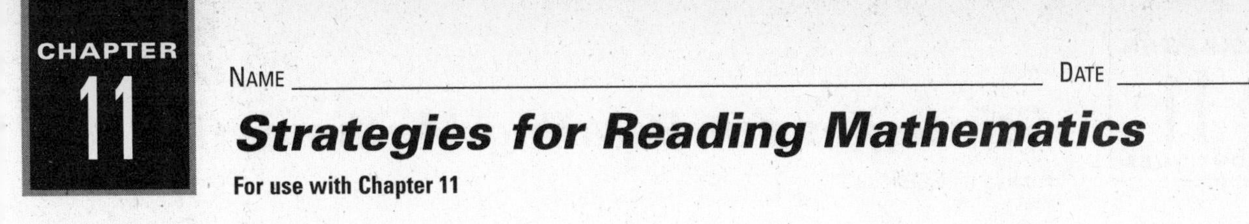

NAME ______________________ DATE __________

Strategies for Reading Mathematics

For use with Chapter 11

Strategy: Reading Formulas

You can use formulas to find the perimeters of polygons, the circumferences of circles, and the areas of polygons and circles. However, before you can use a formula, you need to know how to read it. Since a formula contains variables, you need to know what each variable means and the part of the figure whose measure is represented by the variable.

The area, A, of a rectangle is equal to the length, ℓ, times the width, w.

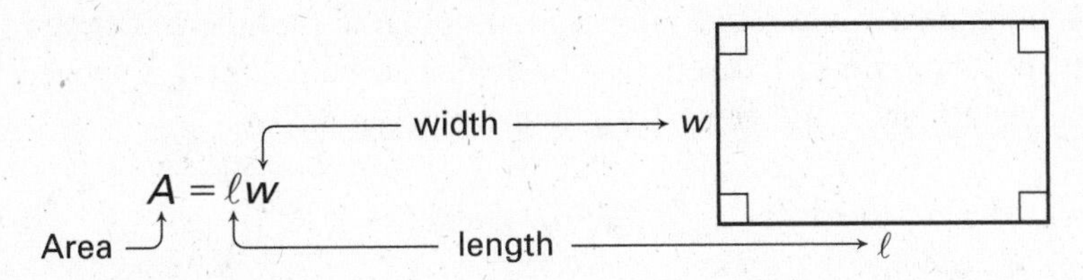

The area, A, of a triangle is equal to one-half the base, b, times the height, h.

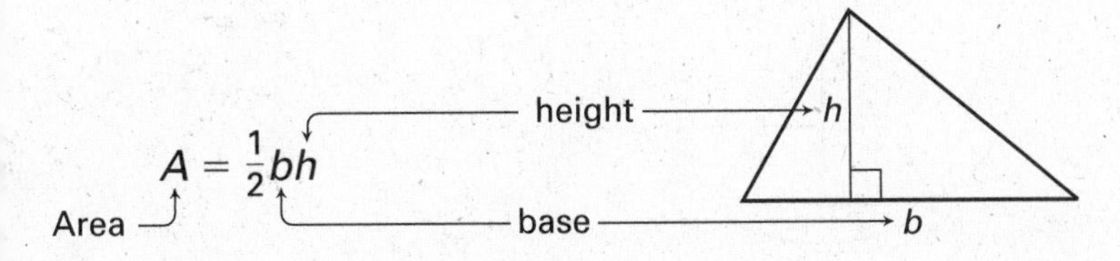

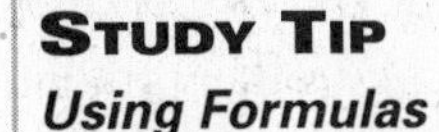

Study Tip

Using Formulas

Before you substitute values into a formula, be sure the units of measure are compatible. For instance, if you are finding the area of a rectangle, both the length and the width must use the same unit of measure. If they do not, convert one measurement so the units are the same.

Questions

1. Why do you think the variable A is used to represent area in both formulas given above? Why do you think the variables ℓ and w are used in the formula for the area of a rectangle? Why do you think b and h are used in the formula for the area of a triangle?
2. Name two things you should know about a variable in a formula.
3. Suppose a triangle has a base of 2 m and a height of 3 m. What are the values of b and h for this triangle? Can you find the area of this triangle using 2 and 3? Why or why not?
4. Suppose a rectangle has a length of 4 ft and a width of 10 in. What are the values of ℓ and w for this rectangle? Can you find the area of this rectangle using 4 and 10? Why or why not?

NAME ______________________________ DATE ____________

Strategies for Reading Mathematics

For use with Chapter 11

Visual Glossary

The Study Guide on page 660 lists the key vocabulary for Chapter 11 as well as review vocabulary from previous chapters. Use the page references on page 660 or the Glossary in the textbook to review key terms from prior chapters. Use the visual glossary below to help you understand some of the key vocabulary in Chapter 11. You may want to copy these diagrams into your notebook and refer to them as you complete the chapter.

Glossary

center of a polygon (p. 670) The center of the polygon's circumscribed circle.

radius of a polygon (p. 670) The radius of the polygon's circumscribed circle.

apothem of a polygon (p. 670) The distance from the center of a polygon to any side of the polygon.

central angle of a regular polygon (p. 671) An angle whose vertex is the center of the polygon and whose sides contain two consecutive vertices of the polygon.

circumference (p. 683) The distance around a circle.

arc length (p. 683) A portion of the circumference of a circle.

sector of a circle (p. 692) The region bounded by two radii of a circle and their intercepted arc.

Parts of a Polygon

To find the center and radius of a regular polygon, find the center and radius of its circumscribed circle. The center of the polygon is needed to find an apothem and central angle of the polygon.

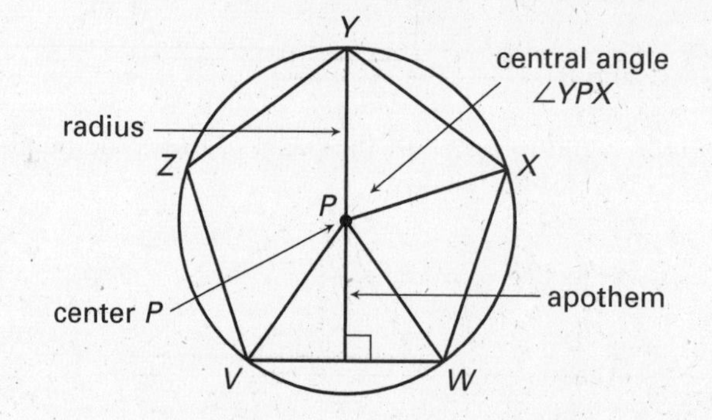

Circumference and Area of a Circle

The length of an arc of a circle is always a portion of the circumference of the circle. The arc length can be used to find the area of a sector of the circle.

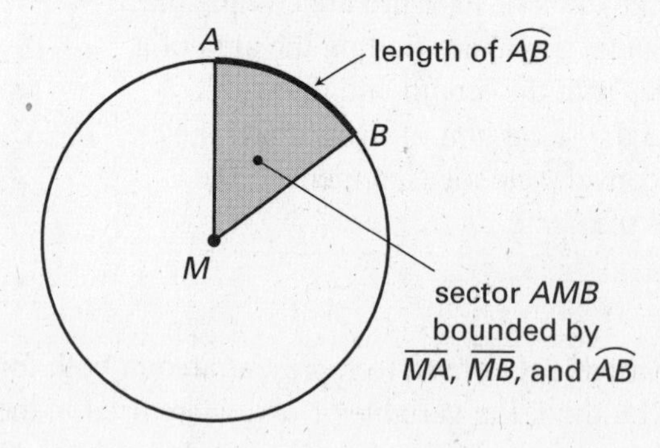

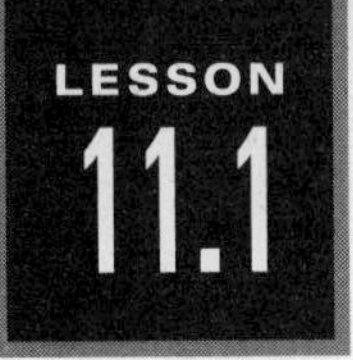

TEACHER'S NAME ________________ CLASS ______ ROOM ______ DATE ______

Lesson Plan

2-day lesson (See *Pacing the Chapter,* TE pages 658C–658D)

For use with pages 661–668

1. Find the measures of interior and exterior angles of polygons.
2. Use measures of angles of polygons to solve real-life problems.

State/Local Objectives ________________________________

____

✓ Check the items you wish to use for this lesson.

STARTING OPTIONS

____ Prerequisite Skills Review: CRB pages 5–6
____ Strategies for Reading Mathematics: CRB pages 7–8
____ Homework Check: TE page 645: Answer Transparencies
____ Warm-Up or Daily Homework Quiz: TE pages 661 and 648, CRB page 11, or Transparencies

TEACHING OPTIONS

____ Motivating the Lesson: TE page 662
____ Lesson Opener (Visual Approach): CRB page 12 or Transparencies
____ Technology Activity with Keystrokes: CRB page 13
____ Examples: Day 1: 1–5, SE pages 662–664; Day 2: See the Extra Examples.
____ Extra Examples: Day 1 or Day 2: 1–5, TE pages 662–664 or Transp.; Internet
____ Closure Question: TE page 664
____ Guided Practice: SE page 665 Day 1: Exs. 1–5; Day 2: See Checkpoint Exs. TE pages 662–664

APPLY/HOMEWORK

Homework Assignment

____ Basic Day 1: 6–40 even; Day 2: 7–41 odd, 49–56, 58–61, 63–73 odd
____ Average Day 1: 6–40 even; Day 2: 7–43 odd, 49–61, 63–73 odd
____ Advanced Day 1: 6–40 even; Day 2: 7–41 odd, 43–45, 49–62, 63–73 odd

Reteaching the Lesson

____ Practice Masters: CRB pages 14–16 (Level A, Level B, Level C)
____ Reteaching with Practice: CRB pages 17–18 or Practice Workbook with Examples
____ Personal Student Tutor

Extending the Lesson

____ Cooperative Learning Activity: CRB page 20
____ Applications (Interdisciplinary): CRB page 21
____ Challenge: SE page 668; CRB page 22 or Internet

ASSESSMENT OPTIONS

____ Checkpoint Exercises: Day 1 or Day 2: TE pages 662–664 or Transp.
____ Daily Homework Quiz (11.1): TE page 668, CRB page 25, or Transparencies
____ Standardized Test Practice: SE page 668; TE page 668; STP Workbook; Transparencies

Notes ________________________________

____

____

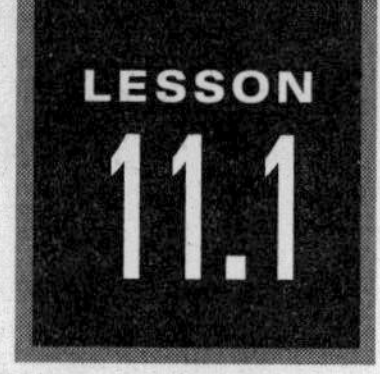

TEACHER'S NAME ______________________ CLASS ______ ROOM ______ DATE ______

Lesson Plan for Block Scheduling

1-day lesson (See *Pacing the Chapter,* TE pages 658C–658D) For use with pages 661–668

1. Find the measures of interior and exterior angles of polygons.
2. Use measures of angles of polygons to solve real-life problems.

State/Local Objectives ______________________

CHAPTER PACING GUIDE	
Day	**Lesson**
1	**11.1 (all)**
2	11.2 (all)
3	11.3 (all)
4	11.4 (all)
5	11.5 (all); 11.6 (begin)
6	11.6 (end); Review Ch. 11
7	Assess Ch. 11; 12.1 (all)

✓ **Check the items you wish to use for this lesson.**

STARTING OPTIONS

____ Prerequisite Skills Review: CRB pages 5–6
____ Strategies for Reading Mathematics: CRB pages 7–8
____ Homework Check: TE page 645: Answer Transparencies
____ Warm-Up or Daily Homework Quiz: TE pages 661 and 648, CRB page 11, or Transparencies

TEACHING OPTIONS

____ Motivating the Lesson: TE page 662
____ Lesson Opener (Visual Approach): CRB page 12 or Transparencies
____ Technology Activity with Keystrokes: CRB page 13
____ Examples 1–5: SE pages 662–664
____ Extra Examples: TE pages 662–664 or Transparencies; Internet
____ Closure Question: TE page 664
____ Guided Practice Exercises: SE page 665

APPLY/HOMEWORK

Homework Assignment

____ Block Schedule: 6–41, 43, 49–61, 63–73 odd

Reteaching the Lesson

____ Practice Masters: CRB pages 14–16 (Level A, Level B, Level C)
____ Reteaching with Practice: CRB pages 17–18 or Practice Workbook with Examples
____ Personal Student Tutor

Extending the Lesson

____ Cooperative Learning Activity: CRB page 20
____ Applications (Interdisciplinary): CRB page 21
____ Challenge: SE page 668; CRB page 22 or Internet

ASSESSMENT OPTIONS

____ Checkpoint Exercises: TE pages 662–664 or Transparencies
____ Daily Homework Quiz (11.1): TE page 668, CRB page 25, or Transparencies
____ Standardized Test Practice: SE page 668; TE page 668; STP Workbook; Transparencies

Notes ______________________

NAME ______________________ DATE __________

Available as a transparency

WARM-UP EXERCISES

For use before Lesson 11.1, pages 661–668

Given the following angle measures in $\triangle ABC$, find $m\angle C$.

1. $m\angle A = 28°, m\angle B = 112°$

2. $m\angle A = 42°, m\angle B = 90°$

3. $m\angle A = 60°, m\angle B = 60°$

4. What is a name for a regular polygon that is also a triangle?

DAILY HOMEWORK QUIZ

For use after Lesson 10.7, pages 641–648

Draw the figure. Then sketch and describe the locus of points on the paper that satisfy the given condition.

1. Line f, the locus of points that are no more than 1 inch from f

2. Point p, the locus of points that are 1 cm or more from p

3. Points $A(1, 2)$ and $B(3, -2)$, the locus of points equidistant from A and B

NAME ______________________ DATE __________

Available as a transparency

Visual Approach Lesson Opener

For use with pages 661–668

Write the name of each regular polygon. Find the measure of one interior angle of the polygon. Use the dashed lines to help. (*Hint:* The sum of the angles that meet at the center of the polygon is 360°. Each triangle formed by dashed lines is isosceles.)

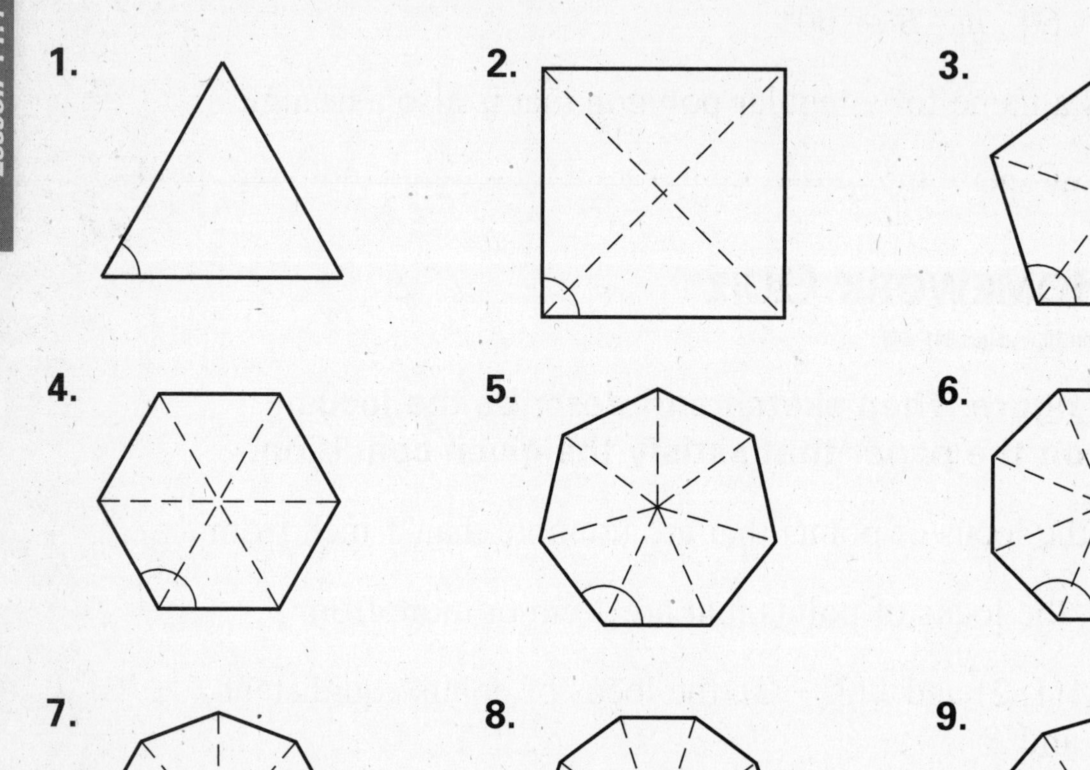

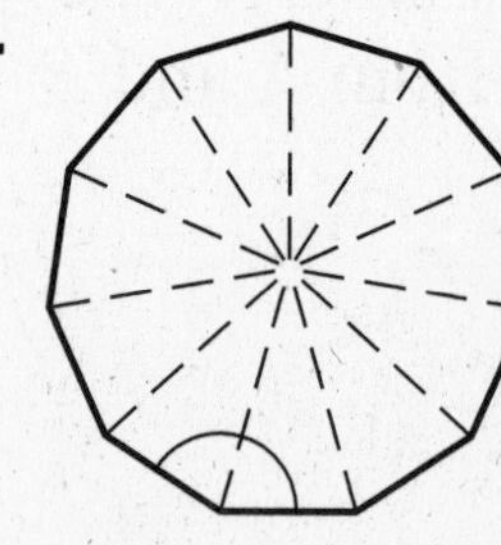

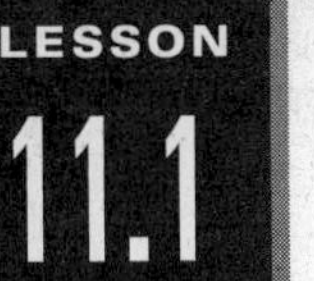

LESSON 11.1

NAME ______________________ DATE __________

Technology Activity Keystrokes

For use with page 667

Keystrokes for Exercises 47 and 48

TI-92

1. Draw a polygon using the polygon command (F3 4).
2. Draw a line through the endpoints of each side using the line command (F2 4).
3. Measure the exterior angles (F6 3).
4. Use the calculator command (F6 6) to verify that the sum of the measures is 180°.
5. Drag a vertex and notice the results.

   F1 1 (Move cursor to a vertex.) ENTER (Use the drag key and the cursor pad to drag the vertex.)

SKETCHPAD

1. Draw a polygon using the segment straightedge tool.
2. Draw a line through the endpoints of each side using the line straightedge tool. Use the point tool to plot a point on each line but not on the polygon.
3. Measure the exterior angles.
4. Choose **Calculate . . .** from the **Measure** menu. Verify that the sum of the measures is 180°.
5. Use the translate selection arrow tool to drag a vertex.

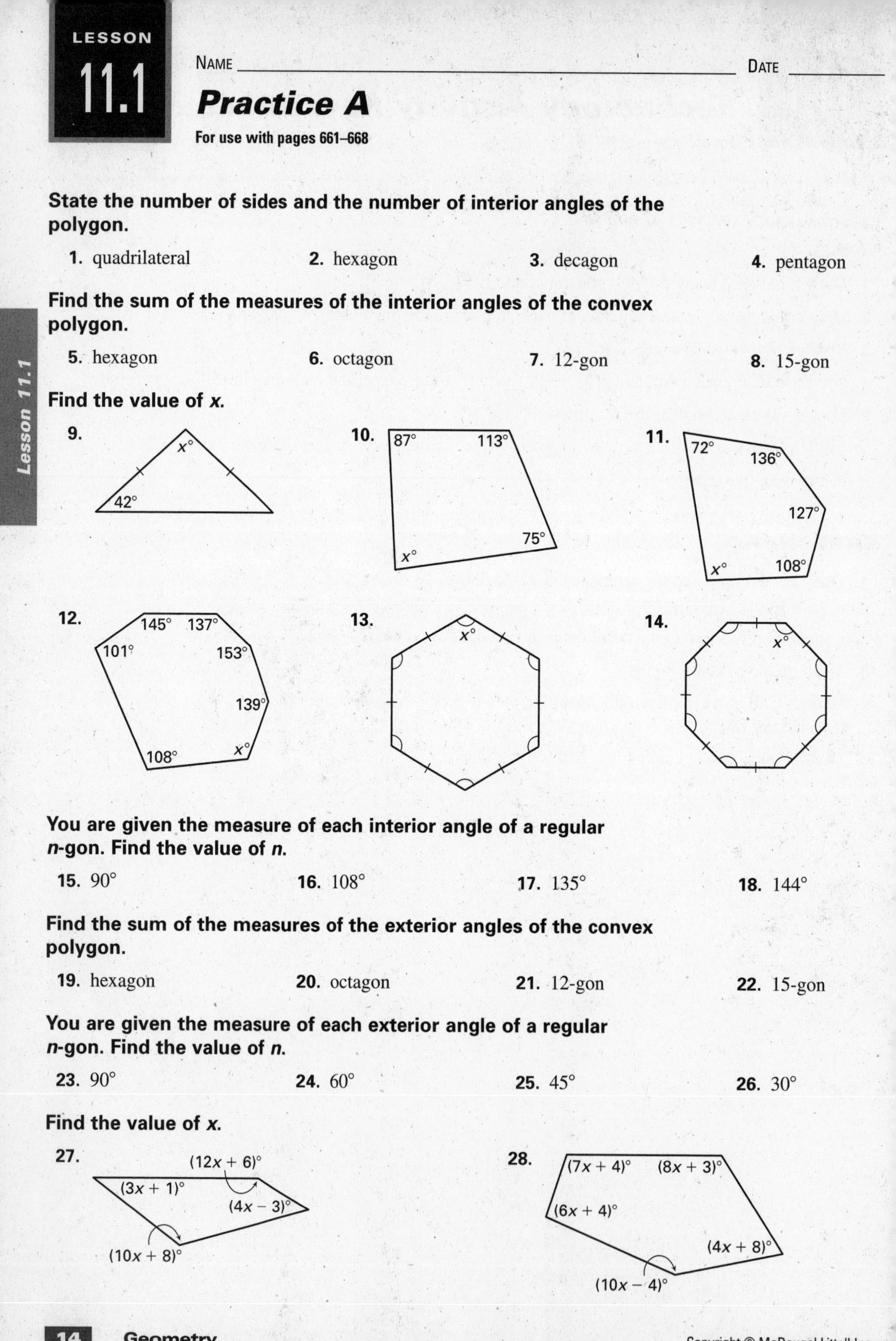

LESSON 11.1

NAME ______________________ DATE __________

Practice A

For use with pages 661–668

State the number of sides and the number of interior angles of the polygon.

1. quadrilateral
2. hexagon
3. decagon
4. pentagon

Find the sum of the measures of the interior angles of the convex polygon.

5. hexagon
6. octagon
7. 12-gon
8. 15-gon

Find the value of x.

9.

10.

11.

12.

13.

14.

You are given the measure of each interior angle of a regular n-gon. Find the value of n.

15. 90°
16. 108°
17. 135°
18. 144°

Find the sum of the measures of the exterior angles of the convex polygon.

19. hexagon
20. octagon
21. 12-gon
22. 15-gon

You are given the measure of each exterior angle of a regular n-gon. Find the value of n.

23. 90°
24. 60°
25. 45°
26. 30°

Find the value of x.

27.

28.

Lesson 11.1

LESSON 11.1

NAME ______________________ DATE __________

Reteaching with Practice

For use with pages 661–668

GOAL **Find the measures of interior and exterior angles of polygons**

VOCABULARY

Theorem 11.1 Polygon Interior Angles Theorem
The sum of the measures of the interior angles of a convex n-gon is $(n - 2) \cdot 180°$.

Corollary to Theorem 11.1
The measure of each interior angle of a regular n-gon is

$$\frac{1}{n} \cdot (n - 2) \cdot 180°, \text{ or } \frac{(n - 2) \cdot 180°}{n}.$$

Theorem 11.2 Polygon Exterior Angles Theorem
The sum of the measures of the exterior angles of a convex polygon, one angle at each vertex, is 360°.

Corollary to Theorem 11.2
The measure of each exterior angle of a regular n-gon is

$$\frac{1}{n} \cdot 360°, \text{ or } \frac{360°}{n}.$$

EXAMPLE 1 ***Finding Measures of Interior Angles of Polygons***

Find the value of x.

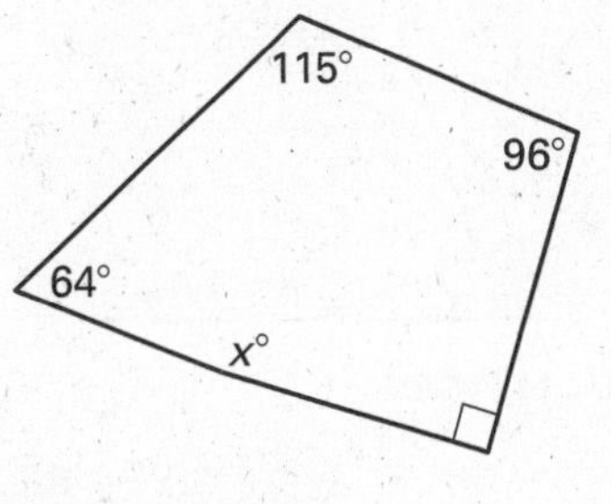

SOLUTION

The sum of the measure of the interior angles of any pentagon is $(5 - 2) \cdot 180° = 3 \cdot 180° = 540°$.

Add the measures of the interior angles of the pentagon.

$64° + 115° + 96° + 90° + x° = 540°$	The sum is 540°.
$365 + x = 540$	Simplify.
$x = 175$	Subtract 365 from each side.

Exercises for Example 1

In Exercises 1–3, find the value of *x*.

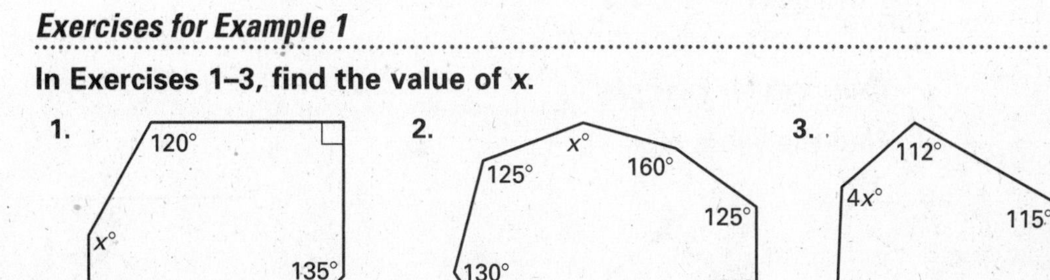

NAME ______________________ DATE  ________

Reteaching with Practice

For use with pages 661–668

EXAMPLE 2 *Finding the Number of Sides of a Polygon*

The measure of each interior angle of a regular polygon is 144°. How many sides does the polygon have?

SOLUTION

$\frac{1}{n} \cdot (n - 2) \cdot 180° = 144°$	Corollary to Theorem 11.1
$(n - 2) \cdot 180 = 144n$	Multiply each side by n.
$180n - 360 = 144n$	Distributive property
$n = 10$	Solve for n.

Exercise for Example 2

4. The measure of each interior angle of a regular n-gon is 156°. Find the value of n.

EXAMPLE 3 *Finding the Measure of an Exterior Angle*

Find the value of x in each diagram.

a.

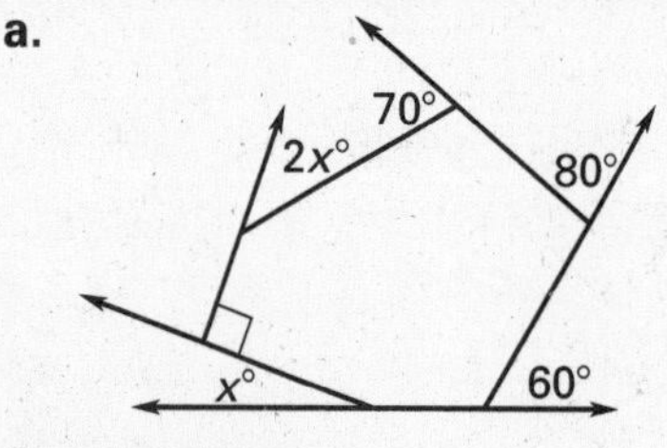

b.

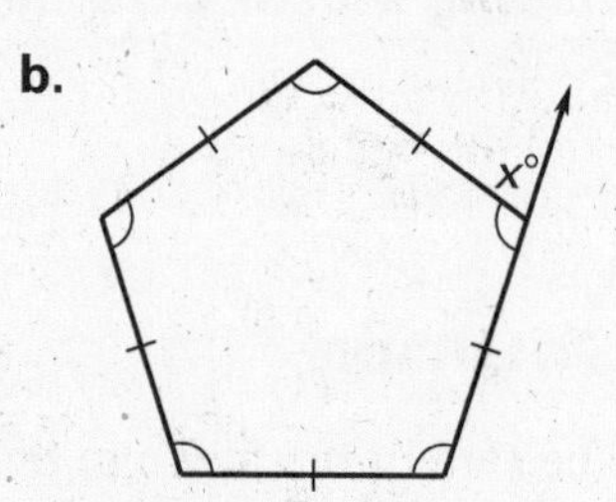

SOLUTION

a. $x° + 90° + 2x° + 70° + 80° + 60° = 360°$	Theorem 11.2
$3x = 60$	Combine like terms.
$x = 20$	Divide each side by 3.

b. $x° = \frac{1}{5} \cdot 360°$	Use $n = 5$ in the Corollary to Theorem 11.2.
$x = 72$	Simplify.

Exercises for Example 3

Find the value of x.

5.

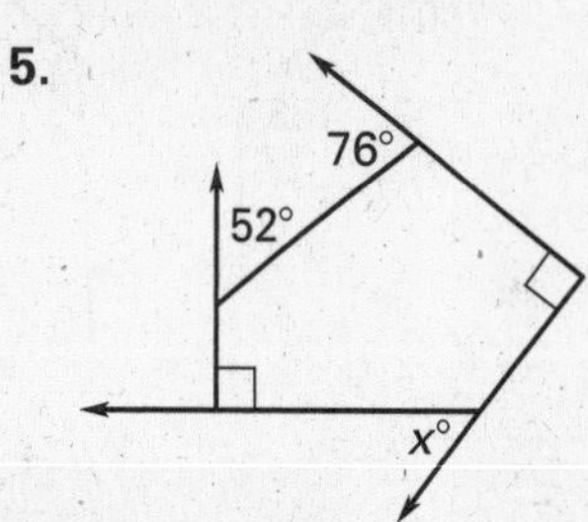

6.

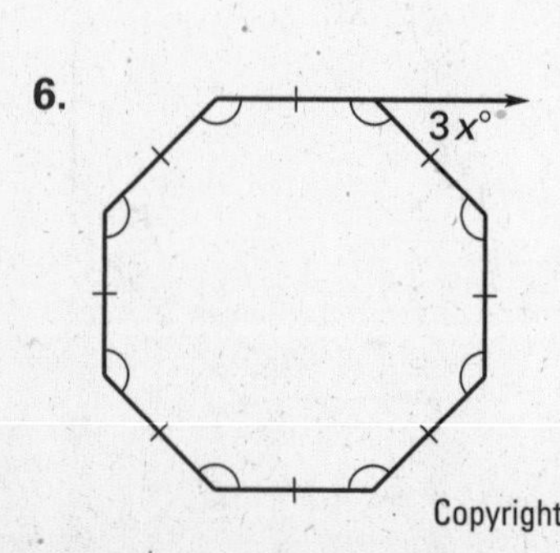

NAME ______________________________ DATE __________

Quick Catch-Up for Absent Students

For use with pages 661–668

The items checked below were covered in class on (date missed) ___________

Lesson 11.1: Angle Measures in Polygons

____ **Goal 1:** Find the measures of interior and exterior angles of polygons. (pp. 661–663)

Material Covered:

____ Activity: Investigating the Sum of Polygon Angle Measures

____ Student Help: Look Back

____ Example 1: Finding Measures of Interior Angles of Polygons

____ Example 2: Finding the Number of Sides of a Polygon

____ Example 3: Finding the Measure of an Exterior Angle

____ **Goal 2:** Use measures of angles of polygons to solve real-life problems. (p. 664)

Material Covered:

____ Example 4: Finding Angle Measures of a Polygon

____ Example 5: Using Angle Measures of a Regular Polygon

____ Other (specify) ______________________________

____

Homework and Additional Learning Support

____ Textbook (specify) pp. 665–668

____

____ Internet: Extra Examples at www.mcdougallittell.com

____ *Reteaching with Practice* worksheet (specify exercises) ______________

____ *Personal Student Tutor* for Lesson 11.1

Cooperative Learning Activity

For use with pages 661–668

 **To investigate the sum of the measures of the interior angles in two different quadrilaterals**

Materials: ruler, protractor, scissors, paper, pencil

Exploring Angle Measures in Polygons

Quadrilaterals are polygons with four sides and four interior angles. There are many different types of quadrilaterals including squares, parallelograms and rectangles. One common property in all quadrilaterals is the sum of the measure of the interior angles.

Instructions

1. On a piece of paper, construct two large, different quadrilaterals.
2. Use a protractor to find the sum of the four angles in each quadrilateral.
3. Cut out the two quadrilaterals using the scissors.
4. Tear off the four angles of the quadrilateral and arrange them around a set point as shown in the diagrams below.

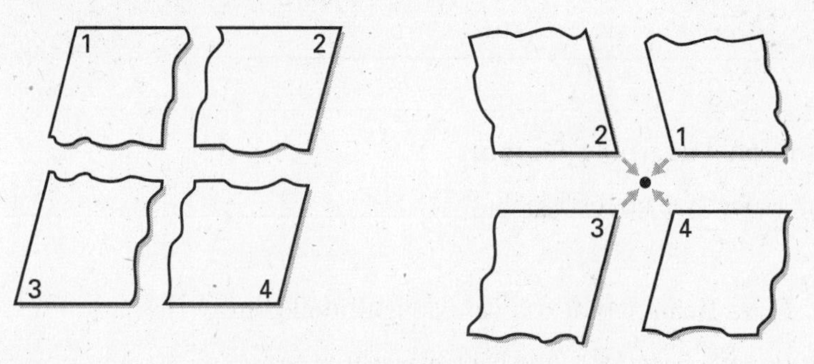

5. Determine the sum of the angles around the point.
6. Repeat steps 1–5 for the second quadrilateral.

Analyzing the Results

1. What is the sum of the angles of each quadrilateral using the protractor?
2. What is the sum of the angles of the quadrilateral found by arranging them around a set point?
3. Are the results the same for each method of measuring the sum of the angles?

NAME ______________________________ DATE ____________

Interdisciplinary Application

For use with pages 661–668

Blue Ash Bicentennial Veterans Memorial

HISTORY The Blue Ash Bicentennial Veterans Memorial located in Blue Ash, Ohio is a very unique memorial. The memorial contains several life-size sculptures, each representing one of the major wars involving the United States: the Revolutionary War, War of 1812, Mexican War, Civil War, Spanish-American War, World Wars I and II, Korean War, Vietnam War, and the Persian Gulf War. The memorial is intended to portray the patriotism and determination of the American spirit, and to inspire all those who see it.

The sculptures are in a 100-foot diameter circle. In the center of the circle, a bronze flagpole is mounted upon a 10-foot wide black granite base in the shape of a nonagon. A cross section of the base is shown below.

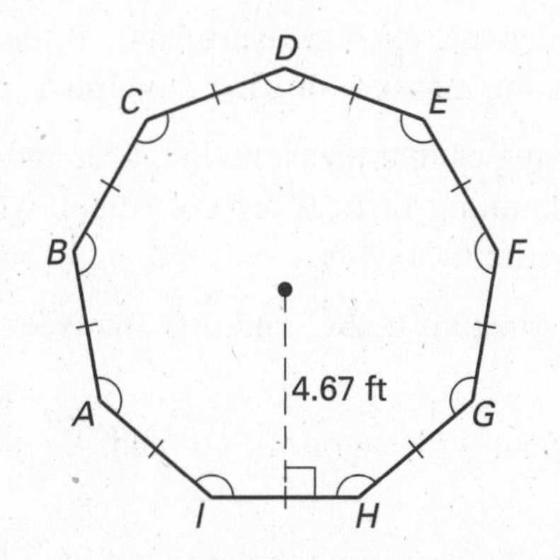

1. Find the measure of each interior angle of nonagon *ABCDEFGHI*.
2. Find the measure of each exterior angle of nonagon *ABCDEFGHI*.
3. Each side of the nonagon measures approximately 3.4 feet in length. What is its perimeter?
4. Using the information from the diagram, determine the area of nonagon *ABCDEFGHI*.

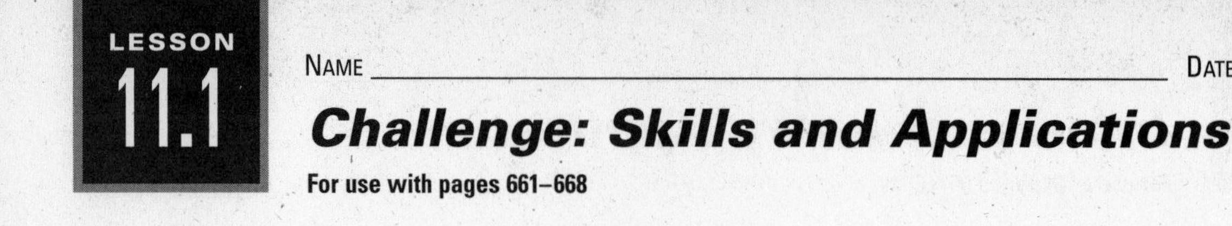

NAME ______________________ DATE ________

Challenge: Skills and Applications

For use with pages 661–668

In Exercises 1–3, find the possible values of x.

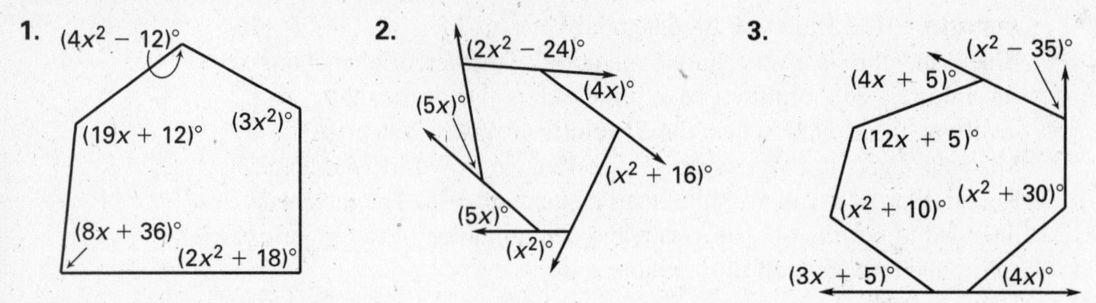

4. A convex heptagon has four interior angles that measure 95°, 118°, 146°, and 160°, respectively. If the remaining interior angles are congruent, what is the measure of each remaining interior angle?

5. A convex 14-gon has six interior angles that each measure 164°. If the remaining interior angles are congruent, what is the measure of each remaining interior angle?

6. A convex octagon has three exterior angles that measure 45°, 65°, and 70°, respectively. If the exterior angles at the remaining vertices are congruent, what is the measure of each of these remaining exterior angles?

7. Polygon *ABCDEFGHI* is a regular nonagon. If $\overleftrightarrow{BC}$ and $\overleftrightarrow{FG}$ intersect at point K, find $m\angle BKF$.

8. Polygon *OPQRSTUVWXYZ* is a regular dodecagon. If $\overleftrightarrow{QR}$ and $\overleftrightarrow{XY}$ intersect at point J, find $m\angle QJK$.

9. Polygon *EFGHIJKLMN* is a regular decagon. Find the measure of $\angle GJM$, an angle formed by two of the diagonals.

In Exercises 10–13, determine whether the figure must be a regular polygon. If so, write the key steps involved in proving it; otherwise, sketch or describe a counterexample.

10. *ABCDEF* is a hexagon in which all sides are congruent, and $m\angle A = m\angle C = m\angle E = 120°$.

11. *GHIJK* is a pentagon in which all sides are congruent, and $m\angle H = m\angle I = m\angle J = 108°$.

12. P is an n-gon that is inscribed in a circle, and all sides of P are congruent.

13. Q is an n-gon that is inscribed in a circle, and all interior angles of Q are congruent.

LESSON 11.2

TEACHER'S NAME ______________________ CLASS ______ ROOM ______ DATE ______

Lesson Plan

2-day lesson (See *Pacing the Chapter,* TE pages 658C–658D) For use with pages 669–675

GOALS

1. **Find the area of an equilateral triangle.**
2. **Find the area of a regular polygon.**

State/Local Objectives ____

____

✓ Check the items you wish to use for this lesson.

STARTING OPTIONS

____ Homework Check: TE page 665: Answer Transparencies
____ Warm-Up or Daily Homework Quiz: TE pages 669 and 668, CRB page 25, or Transparencies

TEACHING OPTIONS

____ Motivating the Lesson: TE page 670
____ Lesson Opener (Activity): CRB page 26 or Transparencies
____ Technology Activity with Keystrokes: CRB pages 27–29
____ Examples: Day 1: 1–3, SE pages 669–671; Day 2: 4, SE page 671
____ Extra Examples: Day 1: TE pages 670–671 or Transp.; Day 2: TE page 671 or Transp.
____ Closure Question: TE page 671
____ Guided Practice: SE page 672 Day 1: Exs. 1–6; Day 2: Exs. 7–8

APPLY/HOMEWORK

Homework Assignment

____ Basic Day 1: 9–29; Day 2: 30–44, 48–52, 54–64
____ Average Day 1: 9–29; Day 2: 30–52, 54–64
____ Advanced Day 1: 9–29; Day 2: 30–64

Reteaching the Lesson

____ Practice Masters: CRB pages 30–32 (Level A, Level B, Level C)
____ Reteaching with Practice: CRB pages 33–34 or Practice Workbook with Examples
____ Personal Student Tutor

Extending the Lesson

____ Applications (Real-Life): CRB page 36
____ Challenge: SE page 675; CRB page 37 or Internet

ASSESSMENT OPTIONS

____ Checkpoint Exercises: Day 1: TE pages 670–671 or Transp.; Day 2: TE page 671 or Transp.
____ Daily Homework Quiz (11.2): TE page 675, CRB page 40, or Transparencies
____ Standardized Test Practice: SE page 675; TE page 675; STP Workbook; Transparencies

Notes ____

____

____

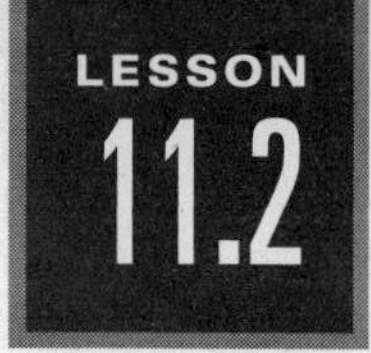

TEACHER'S NAME ______________________ CLASS ________ ROOM ______ DATE ______

Lesson Plan for Block Scheduling

1-day lesson (See *Pacing the Chapter,* TE pages 658C–658D) For use with pages 669–675

1. **Find the area of an equilateral triangle.**
2. **Find the area of a regular polygon.**

State/Local Objectives ______________________

CHAPTER PACING GUIDE	
Day	**Lesson**
1	11.1 (all)
2	**11.2 (all)**
3	11.3 (all)
4	11.4 (all)
5	11.5 (all); 11.6 (begin)
6	11.6 (end); Review Ch. 11
7	Assess Ch. 11; 12.1 (all)

✓ **Check the items you wish to use for this lesson.**

STARTING OPTIONS

____ Homework Check: TE page 665: Answer Transparencies
____ Warm-Up or Daily Homework Quiz: TE pages 669 and 668, CRB page 25, or Transparencies

TEACHING OPTIONS

____ Motivating the Lesson: TE page 670
____ Lesson Opener (Activity): CRB page 26 or Transparencies
____ Technology Activity with Keystrokes: CRB pages 27–29
____ Examples 1–4: SE pages 669–671
____ Extra Examples: TE pages 670–671 or Transparencies
____ Closure Question: TE page 671
____ Guided Practice Exercises: SE page 672

APPLY/HOMEWORK

Homework Assignment

____ Block Schedule: 9–53, 55–66

Reteaching the Lesson

____ Practice Masters: CRB pages 30–32 (Level A, Level B, Level C)
____ Reteaching with Practice: CRB pages 33–34 or Practice Workbook with Examples
____ Personal Student Tutor

Extending the Lesson

____ Applications (Real-Life): CRB page 36
____ Challenge: SE page 675; CRB page 37 or Internet

ASSESSMENT OPTIONS

____ Checkpoint Exercises: TE pages 670–671 or Transparencies
____ Daily Homework Quiz (11.2): TE page 675, CRB page 40, or Transparencies
____ Standardized Test Practice: SE page 675; TE page 675; STP Workbook; Transparencies

Notes ______________________

LESSON 11.2

NAME ______________________ DATE __________

Available as a transparency

WARM-UP EXERCISES

For use before Lesson 11.2, pages 669–675

Use the figure to find each measure.

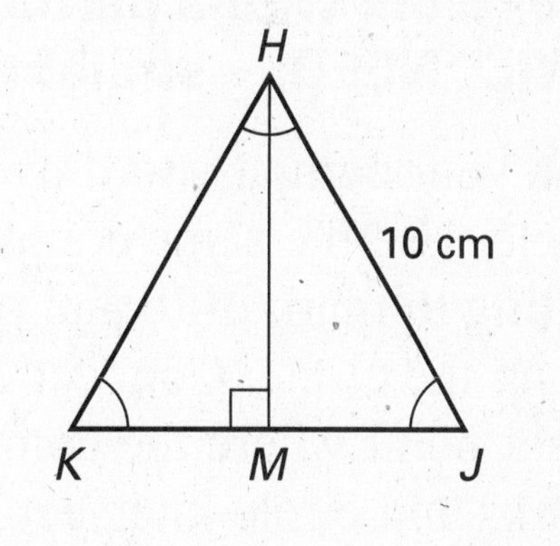

1. KJ

2. $m\angle K$

3. $m\angle KHM$

4. HM

DAILY HOMEWORK QUIZ

For use after Lesson 11.1, pages 661–668

1. Find the sum of the measures of the interior angles in a convex 16-gon.

2. Find the value of x.

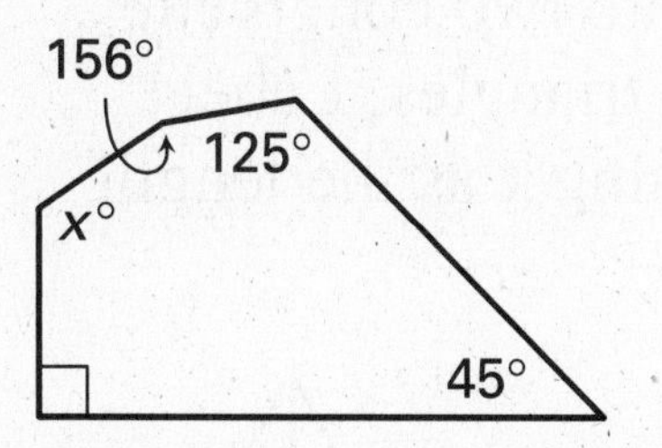

3. The measure of each interior angle of a regular polygon is 162°. How many sides does the polygon have?

4. The measure of each exterior angle of a regular polygon is 6°. How many sides does the polygon have?

Lesson 11.2

NAME ______________________ DATE __________

Available as a transparency

Activity Lesson Opener

For use with pages 669–675

SET UP: Work with a partner.
YOU WILL NEED: • straightedge • compass

1. Use your compass to draw a circle. Use the same compass setting to mark off equal parts along the circle. Connect every other point where the compass marks and circle intersect. You have constructed an equilateral triangle.

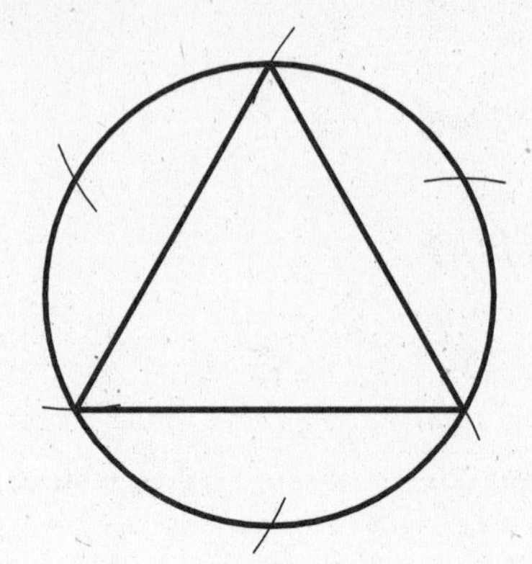

2. Set your compass to be the length of one side of the triangle. Use this compass setting to draw a circle with center at one vertex of the triangle. Draw another circle with the same radius at another vertex of the triangle. Connect the two points of intersection of the circles. You have constructed an altitude of the equilateral triangle.

3. The altitude divides the equilateral triangle into two congruent triangles. Label the angle measures of all the triangles. Label the lengths of the sides of all the triangles, using x as the length of a side of the equilateral triangle.

4. You can use the base length b and height h to find the area of any triangle.

Use the formula Area $= \frac{1}{2}bh$ to find the area of the equilateral triangle you constructed. You now have a formula for the area of an equilateral triangle in terms of x, the length of a side.

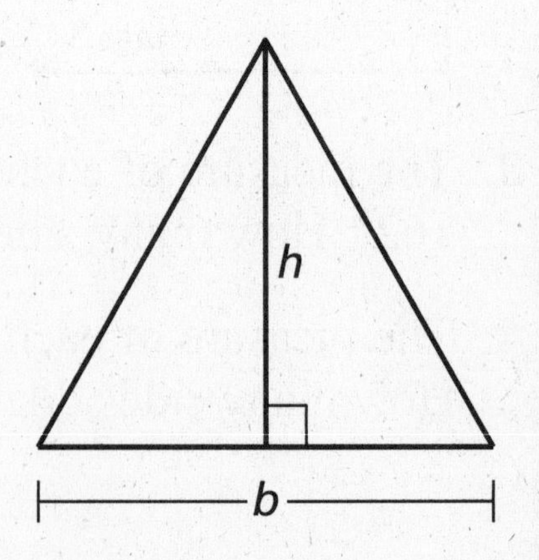

NAME ______________________ DATE __________

Technology Activity

For use with pages 669–675

GOAL **To determine a formula for finding the area of a regular hexagon.**

In this activity, you will develop a formula for the area of a regular hexagon by dividing the hexagon into congruent equilateral triangles.

Activity

1. Construct a regular hexagon.
2. Draw a segment from the hexagon's center to each vertex.
3. Use the triangles formed as a result of Step 2 to calculate the area of the hexagon.

Exercises

1. Write a formula to find the area of a regular hexagon based on dividing it into congruent equilateral triangles.
2. Find the area of a regular hexagon if the length of one side is 4 units.
3. Describe how you can use what you learned in this activity to find the area of a regular hexagon.

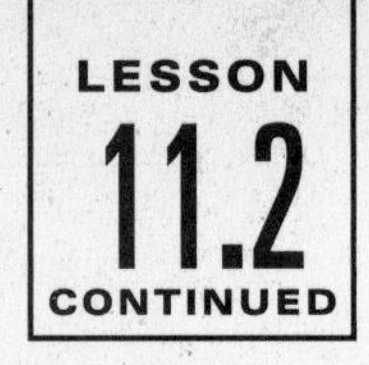

NAME _______________ DATE _______________

Technology Activity Keystrokes

For use with pages 669–675

TI-92

1. Construct a regular hexagon.

   F3 5 (Place cursor in center of screen.) ENTER (Move cursor for desired size of hexagon.) ENTER (Move cursor until number of sides indicated is 6.) ENTER

2. Draw a segment from the hexagon's center to each vertex using the segment command (F2 5).

3. Calculate the area of the hexagon by finding the area of one of the triangles and multiplying by 6. Use the fact that each triangle is equilateral and the area of an equilateral triangle is $A = \frac{1}{4}\sqrt{3}s^2$. Use the distance and length command to measure a side (F6 1). Calculate the area of the equilateral triangle and multiply the area by 6 to obtain the hexagon's area.

Lesson 11.2

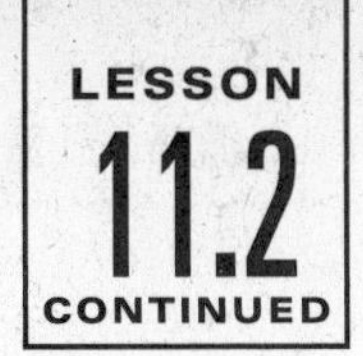

NAME ______________________________ DATE ____________

Technology Activity Keystrokes

For use with pages 669–675

SKETCHPAD

1. Draw a circle. Draw $\overline{AB}$, a radius of the circle. Construct a circle with center B and radius AB by selecting point B and $\overline{AB}$ and choose **Circle by Center and Radius** from the **Construct** menu. Construct intersection point C of this circle and the original circle. Construct a circle with center C and radius AB. Construct intersection point D of this circle and the original circle. Construct a circle with center D and radius AB. Construct intersection point E of this circle and the original circle. Construct a circle with center E and radius AB. Construct intersection point F of this circle and the original circle. Construct a circle with center F and radius AB. Construct intersection point G of this circle and the original circle. Use the segment straightedge tool to construct the hexagon's sides ($\overline{BC}$, $\overline{CD}$, and so on). Hide the circles.
2. Draw a segment from the hexagon's center to each vertex using the segment straightedge tool.
3. Calculate the area of the hexagon by finding the area of one of the triangles and multiplying by 6. Use the fact that each triangle is equilateral and the area of an equilateral triangle is $A = \frac{1}{4}\sqrt{3}s^2$.

LESSON 11.2

NAME ______________________ DATE ________

Practice A

For use with pages 669–675

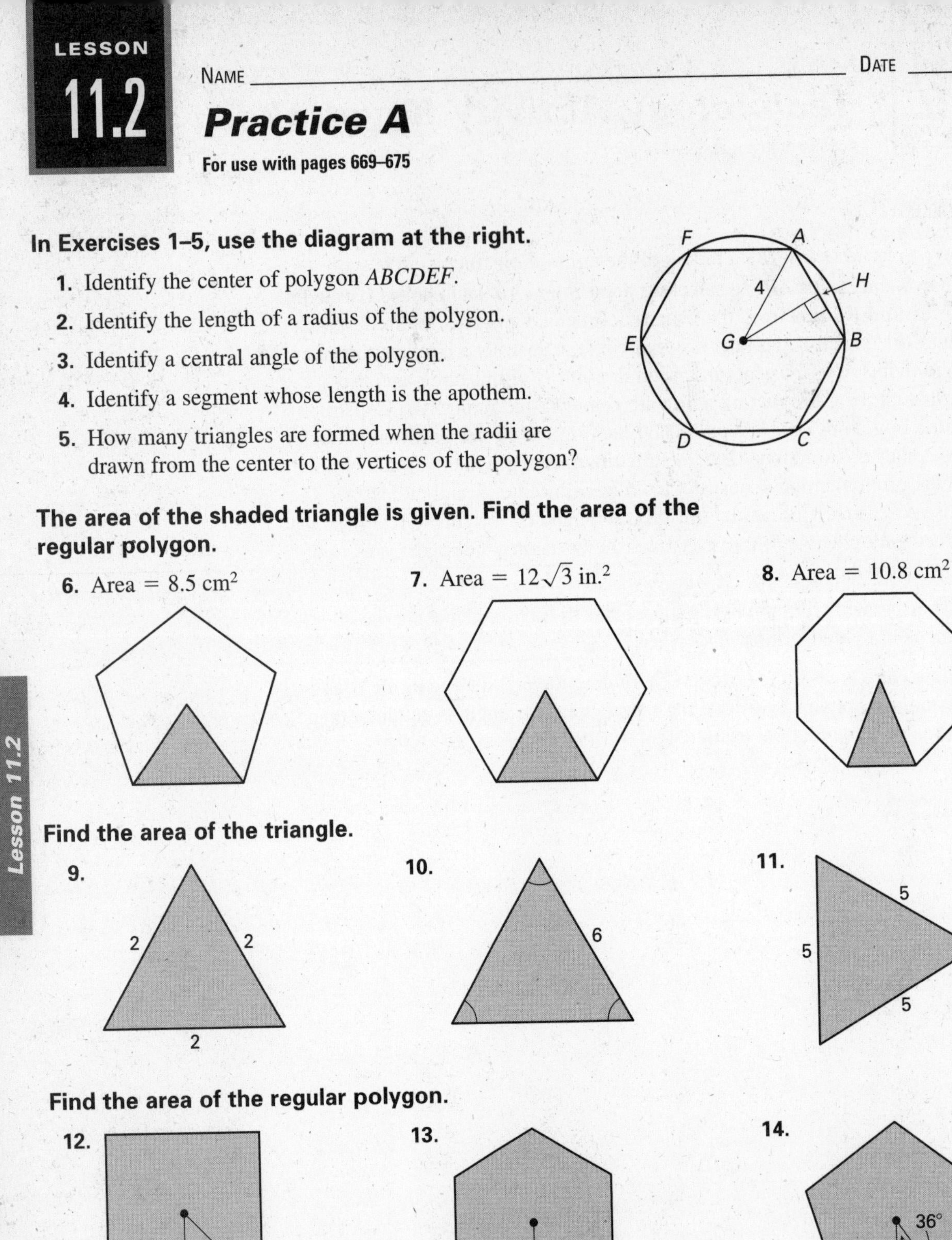

In Exercises 1–5, use the diagram at the right.

1. Identify the center of polygon *ABCDEF*.
2. Identify the length of a radius of the polygon.
3. Identify a central angle of the polygon.
4. Identify a segment whose length is the apothem.
5. How many triangles are formed when the radii are drawn from the center to the vertices of the polygon?

The area of the shaded triangle is given. Find the area of the regular polygon.

6. Area = 8.5 cm^2
7. Area = $12\sqrt{3} \text{ in.}^2$
8. Area = 10.8 cm^2

Find the area of the triangle.

9.

10.

11.

Find the area of the regular polygon.

12.

13.

14.

Find the perimeter and area of the regular polygon.

15.

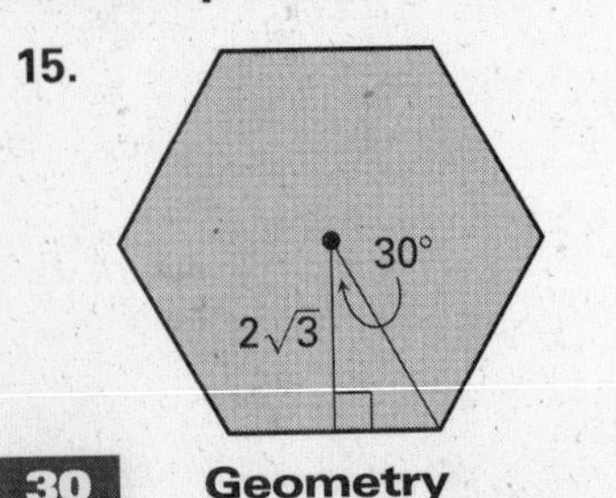

16.

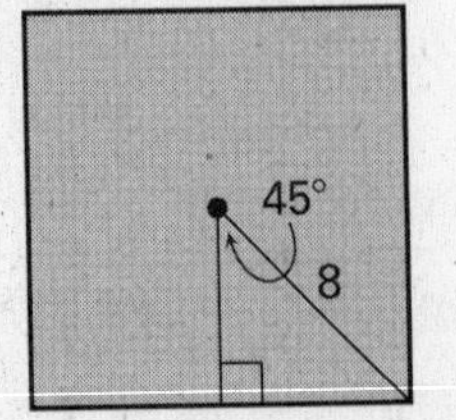

17.

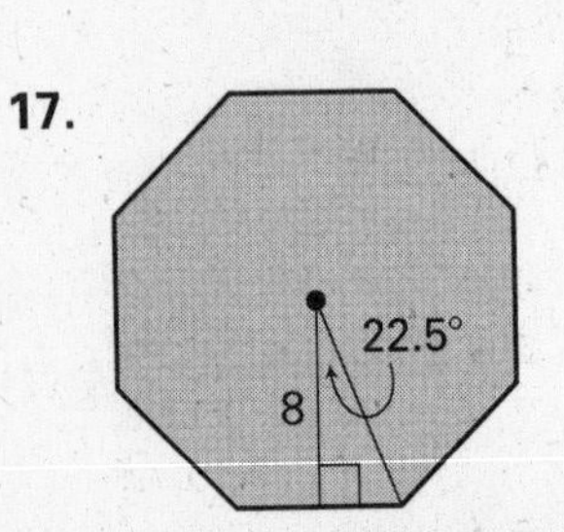

NAME _______________ DATE _______

Practice B

For use with pages 669–675

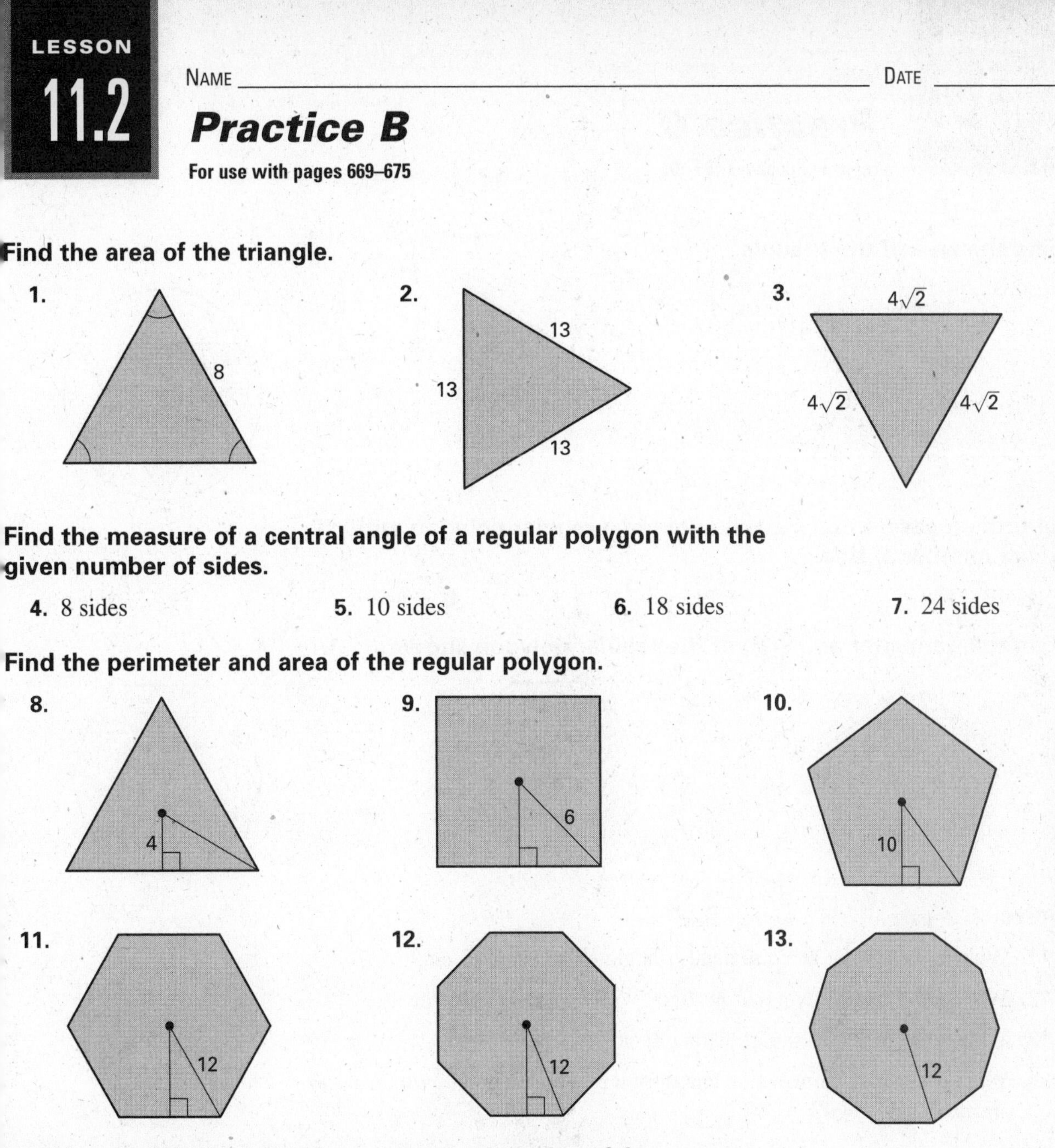

Find the area of the triangle.

1.

2.

3.

Find the measure of a central angle of a regular polygon with the given number of sides.

4. 8 sides
5. 10 sides
6. 18 sides
7. 24 sides

Find the perimeter and area of the regular polygon.

8.

9.

10.

11.

12.

13.

Decide whether the following statement is *true* or *false*.

14. As the number of sides of a regular polygon increases, the measure of the central angle also increases.
15. The radius of a regular polygon is always greater than the apothem.
16. The radius of a regular polygon is always greater than the side length.
17. The area of a regular hexagon of radius 4 is six times greater than the area of a regular triangle of radius 4.

In Exercises 18–20, use the diagram at the right.

18. What is the area of one square?
19. What is the area of one regular octagon?
20. The cost of ceramic tiling is $.025 per square inch. What would be the cost of the design shown?

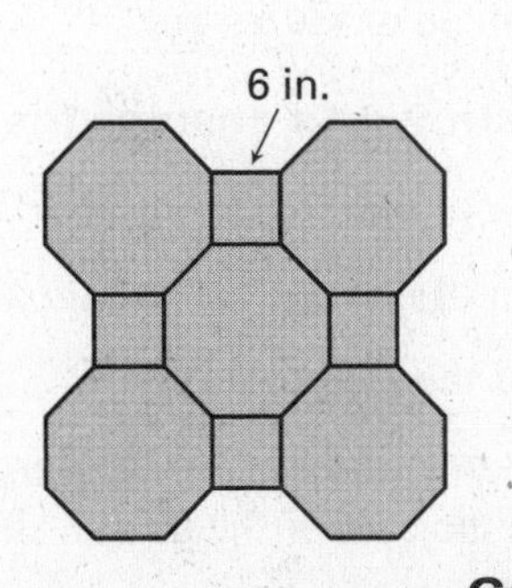

NAME ______________________________ DATE __________

Practice C

For use with pages 669–675

Find the area of the triangle.

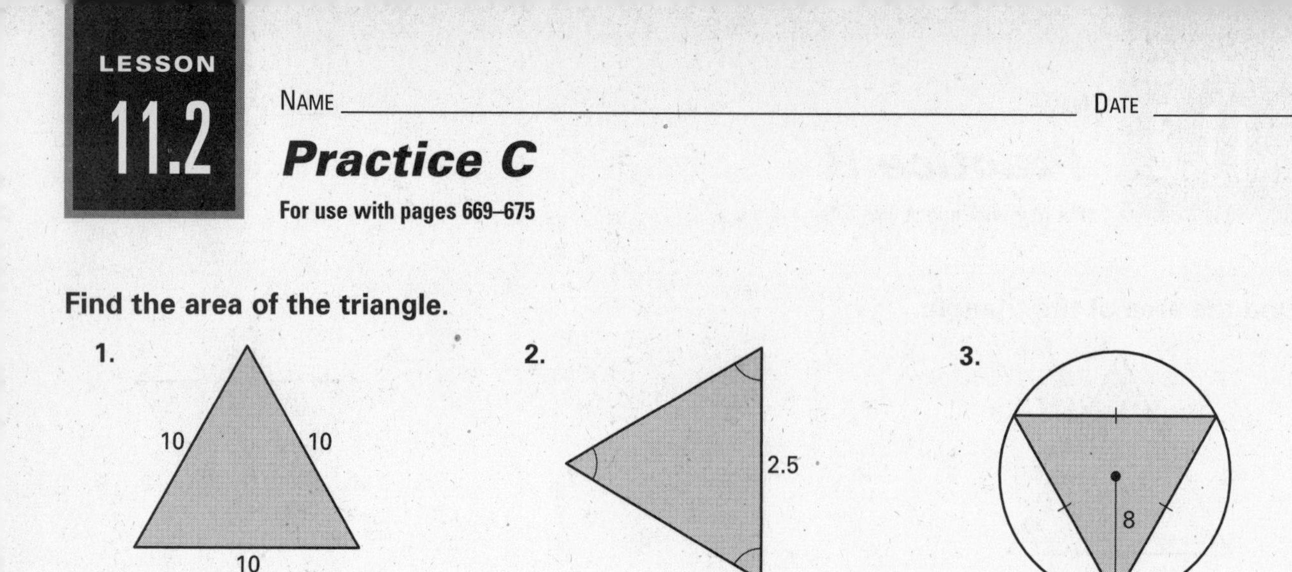

Find the measure of a central angle of a regular polygon with the given number of sides.

4. 12 sides **5.** 15 sides **6.** 25 sides **7.** 32 sides

Find the perimeter and area of the regular polygon shown.

8.

9.

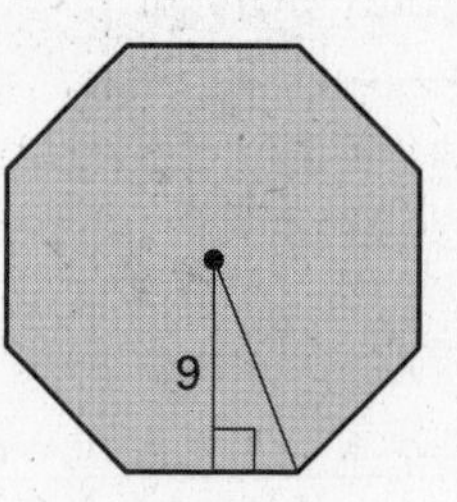

10.

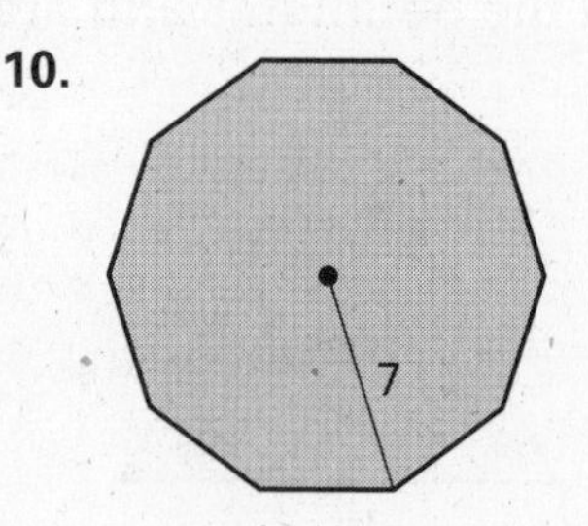

Solve.

11. What is the area of an equilateral triangle with radius 15 cm?

12. What is the area of a regular hexagon with apothem 7.5 inches?

13. What is the area of a square with a diagonal 6.3 cm?

14. What is the approximate side length of a regular hexagon with area 100 square centimeters?

Decide whether the statement is *always, sometimes,* or *never* true.

15. If the number of sides of a regular polygon is n, then the measure of the central angle is $360° \div n$.

16. The radius of a regular polygon is greater than the apothem.

17. The radius of a regular hexagon of radius r is six times greater than the area of a regular triangle of radius r.

18. The area of a regular hexagon of radius r is six times greater than the area of a regular triangle of radius r.

Use the diagram shown to answer the following.

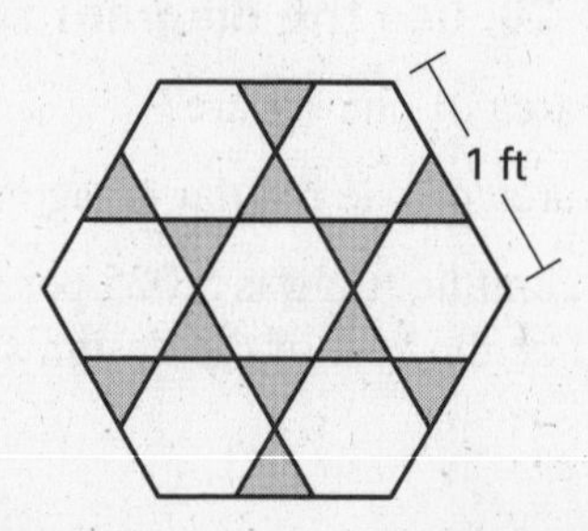

19. What is the area of one regular triangle?

20. What is the area of one regular hexagon?

21. What percent of the tiling is shaded?

22. The cost of ceramic tiling is $0.025 per square inch. What would be the cost of the design shown?

NAME ______________________________ DATE __________

Reteaching with Practice

For use with pages 669–675

GOAL Find the area of an equilateral triangle and a regular polygon

VOCABULARY

The **center of a regular polygon** is the center of its circumscribed circle.

The **radius of a regular polygon** is the radius of its circumscribed circle.

The distance from the center to any side of a regular polygon is called the **apothem of the polygon**.

A **central angle of a regular polygon** is an angle whose vertex is the center and whose sides contain two consecutive vertices of the polygon.

Theorem 11.3 Area of an Equilateral Triangle The area of an equilateral triangle is one fourth the square of the length of the side times $\sqrt{3}$.

$A = \frac{1}{4}\sqrt{3}s^2$

Theorem 11.4 Area of a Regular Polygon The area of a regular n-gon with side length s is half the product of the apothem a and the perimeter P, so $A = \frac{1}{2}aP$, or $A = \frac{1}{2}a \cdot ns$.

EXAMPLE 1 *Finding the Area of an Equilateral Triangle*

Find the area of an equilateral triangle with 4 foot sides.

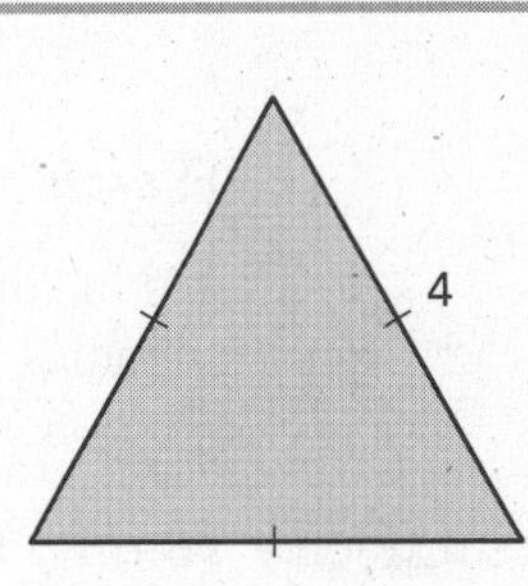

SOLUTION

Use $s = 4$ in the formula of Theorem 11.3.

$$A = \frac{1}{4}\sqrt{3}s^2 = \frac{1}{4}\sqrt{3}(4^2)$$

$$= \frac{1}{4}\sqrt{3}(16) = \frac{1}{4}(16)\sqrt{3} = 4\sqrt{3} \text{ square feet}$$

Using a calculator, the area is about 6.9 square feet.

Exercises for Example 1

Find the area of the triangle.

1. **2.** **3.**

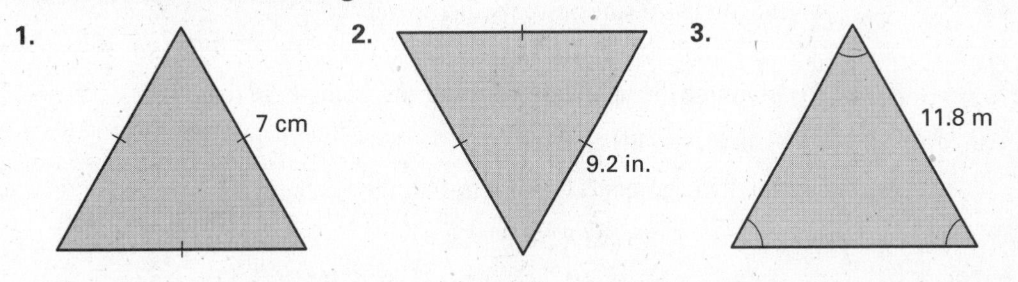

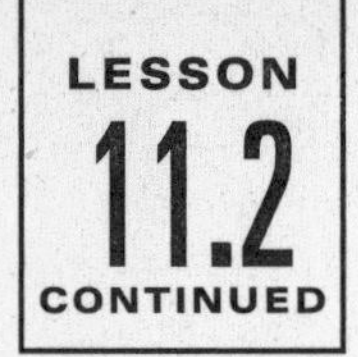

NAME ______________________________ DATE __________

Reteaching with Practice

For use with pages 669–675

EXAMPLE 2 *Finding the Area of a Regular Polygon*

A regular octagon is inscribed in a circle with radius 2 units. Find the area of the octagon.

SOLUTION

To apply the formula for the area of a regular octagon, you must find its apothem and perimeter.

The measure of central $\angle ABC$ is $\frac{1}{8} \cdot 360° = 45°$.

In isosceles triangle ABC, the altitude to base $\overline{AC}$ also bisects $\angle ABC$ and side $\overline{AC}$. The measure of $\angle DBC$ is 22.5°. In $\triangle BDC$, you can use trigonometric ratios to find the lengths of the legs.

$$\cos 22.5° = \frac{BD}{BC} = \frac{BD}{2} \text{ and } \sin 22.5° = \frac{DC}{BC} = \frac{DC}{2}$$

So, the octagon has an apothem of $a = BD = 2 \cdot \cos 22.5°$ and perimeter of $P = 8(AC) = 8(2 \cdot DC) = 8(2 \cdot 2 \cdot \sin 22.5°) = 32 \cdot \sin 22.5°$. The area of the octagon is

$$A = \frac{1}{2}aP = \frac{1}{2}(2 \cdot \cos 22.5°)(32 \cdot \sin 22.5°) \approx 11.3 \text{ square units.}$$

Exercise for Example 2

4. Find the area of a regular pentagon inscribed in a circle with radius 3 units.

EXAMPLE 3 *Finding the Perimeter and Area of a Regular Polygon*

Find the perimeter and area of a regular hexagon with side length of 4 cm and radius 4 cm.

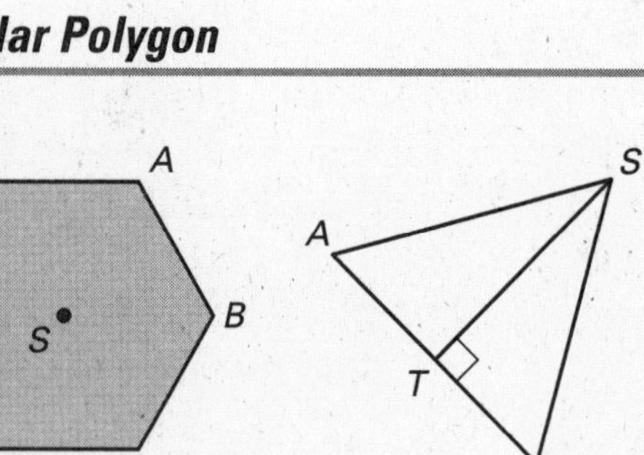

SOLUTION

A hexagon has 6 sides. So, the perimeter is $P = 6(4) = 24$ cm.

To determine the apothem, consider the triangle SBT.

$BT = \frac{1}{2}(BA) = \frac{1}{2}(4) = 2$ cm.

Use the Pythagorean Theorem to find the apothem ST.

$a = \sqrt{4^2 - 2^2} = 2\sqrt{3}$ cm.

So, the area of the hexagon is $A = \frac{1}{2}aP = \frac{1}{2}(2\sqrt{3})(24) = 24\sqrt{3}$ cm^2.

Exercise for Example 3

Find the perimeter and area of the regular polygon described.

5. Regular octagon with side length 9.18 feet and radius 12 feet.

NAME ______________________ DATE ________

Quick Catch-Up for Absent Students

For use with pages 669–675

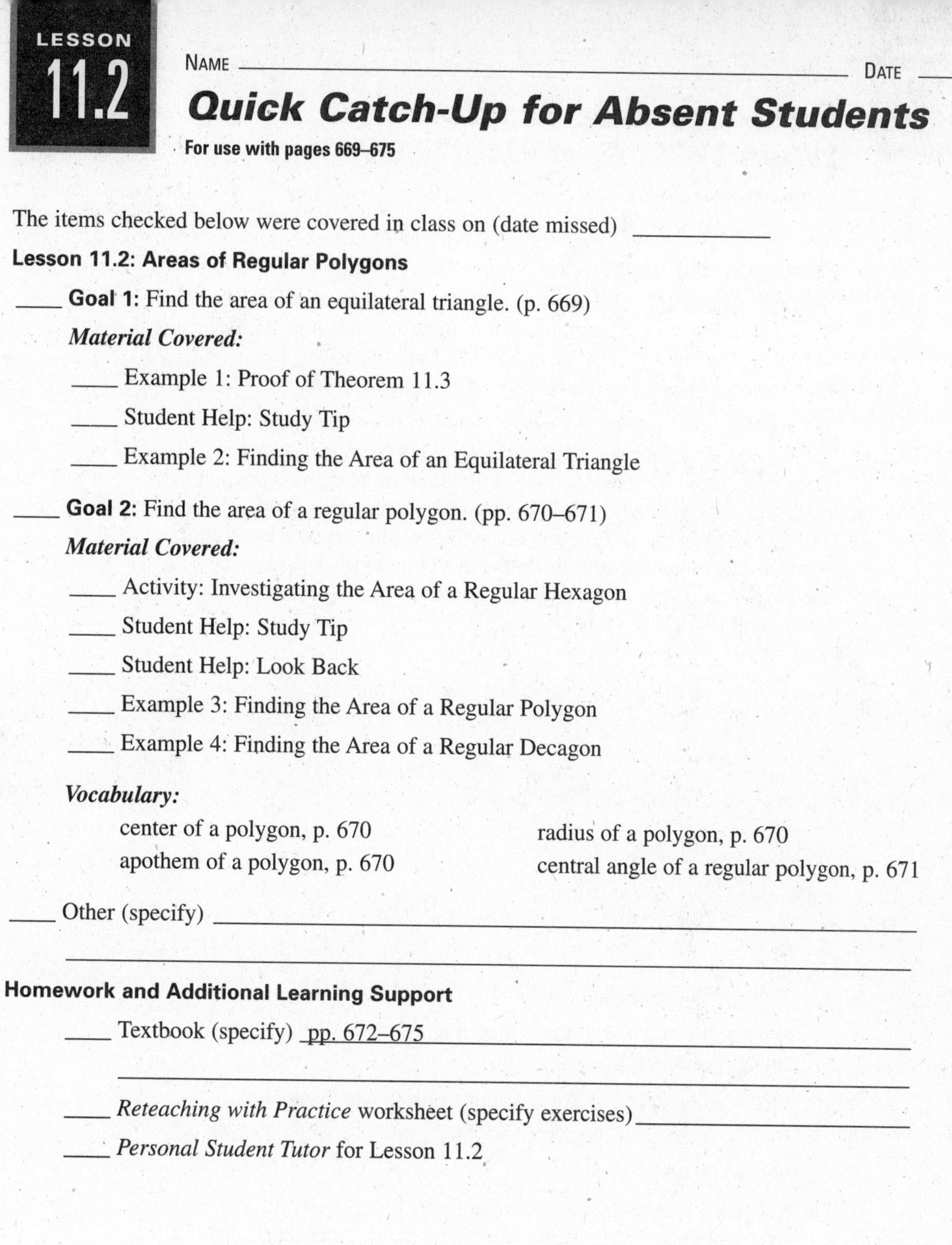

The items checked below were covered in class on (date missed) ____________

Lesson 11.2: Areas of Regular Polygons

____ **Goal 1:** Find the area of an equilateral triangle. (p. 669)

Material Covered:

____ Example 1: Proof of Theorem 11.3

____ Student Help: Study Tip

____ Example 2: Finding the Area of an Equilateral Triangle

____ **Goal 2:** Find the area of a regular polygon. (pp. 670–671)

Material Covered:

____ Activity: Investigating the Area of a Regular Hexagon

____ Student Help: Study Tip

____ Student Help: Look Back

____ Example 3: Finding the Area of a Regular Polygon

____ Example 4: Finding the Area of a Regular Decagon

Vocabulary:

center of a polygon, p. 670
radius of a polygon, p. 670
apothem of a polygon, p. 670
central angle of a regular polygon, p. 671

____ Other (specify) ______________________

____

Homework and Additional Learning Support

____ Textbook (specify) pp. 672–675

____

____ *Reteaching with Practice* worksheet (specify exercises) ____________

____ *Personal Student Tutor* for Lesson 11.2

NAME ______________________ DATE ____________

Real-Life Application: When Will I Ever Use This?

For use with pages 669–675

Pentagon Building

The Pentagon Building, headquarters of the Department of Defense of the United States government, is one of the world's largest office buildings. The building lies just across the Potomac River from Washington, D.C. in Arlington, Virginia.

The building was constructed by army engineers and was completed in 1943. The Pentagon has approximately 23,000 employees and is virtually a city in itself with private restaurants, a bank, heliport, and a television station. It was built in the shape of a pentagon with each outer wall having a length of 921 feet. The polygon *ABCDE* shown below represents the outer edge of the building with point *F* as its center. The smaller pentagon represents the courtyard that lies within the Pentagon Building.

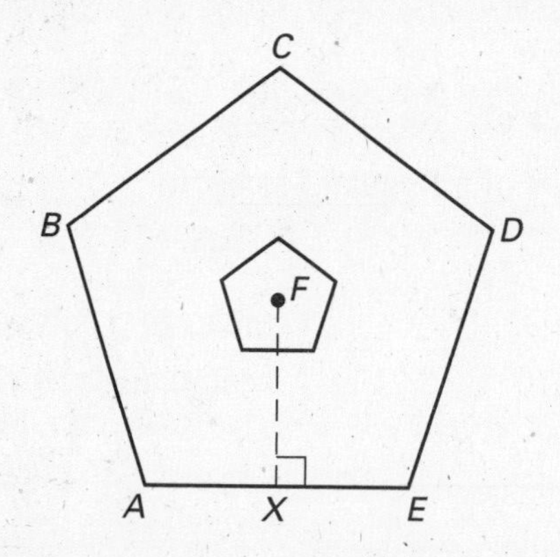

1. What is the perimeter of the Pentagon Building?
2. What is the measure of the central angle $\angle AFE$? What type of triangle is $\triangle AFE$?
3. Find the length of $\overline{XF}$. Round your result to one decimal place.
4. Find the total area of the Pentagon building. Round your result to one decimal place.
5. Suppose the inner pentagon that represents the courtyard has a side length of approximately 307 feet. What is the approximate ratio of its area compared to the area of the entire Pentagon?

NAME ______________________ DATE __________

Challenge: Skills and Applications

For use with pages 669–675

In Exercises 1–8, refer to the diagram. *O* is the center of a regular *n*-gon, and *P* and *Q* are adjacent vertices of the polygon. *M* is the midpoint of $\overline{PQ}$.

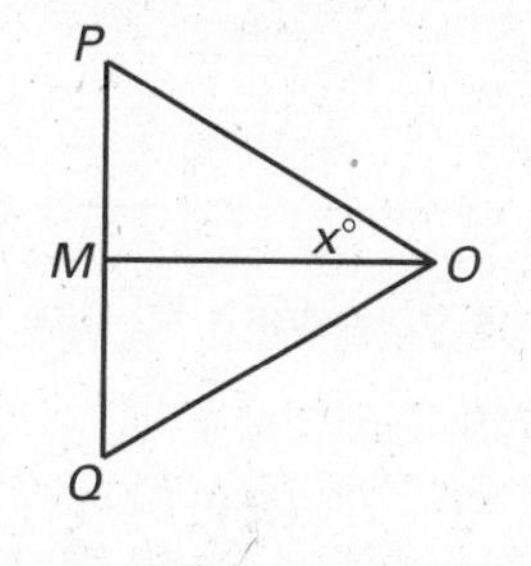

1. Identify each of OM, OP, and PQ as the side length, the radius, or the apothem of the n-gon.
2. Find a formula for x in terms of n.
3. Find a formula for the apothem length a in terms of n and the radius r. (*Hint:* Use your answer to Exercise 2.)
4. Find a formula for the side length s in terms of n and the radius r.
5. Find a formula for the apothem a in terms of n and the side length s.
6. Find a formula for the area of a regular n-gon in terms of n and the side length s.
7. Find a formula for the area of a regular n-gon in terms of n and the apothem length a.
8. Find a formula for the area of a regular n-gon in terms of n and the radius r.
9. Consider a regular n-gon inscribed in a circle of radius 1. Use a calculator and the result of Exercise 8 to find the area of the n-gon for $n = 4$, 8, 25, 50, and 100. What number does the area seem to approach as n increases? Round decimals to the nearest hundredth.

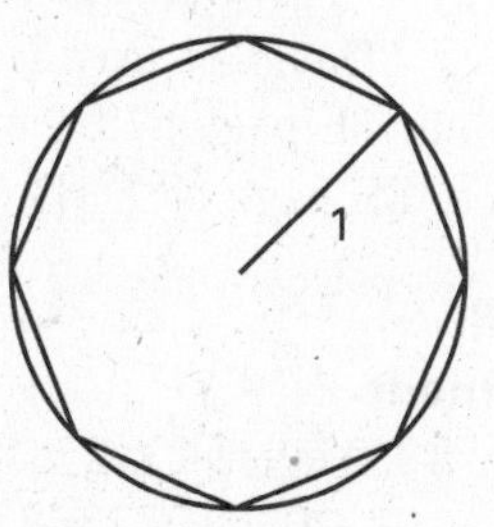

10. Refer to the diagram, which shows an arbitrary point P inside a regular pentagon, along with perpendiculars drawn from P to the sides of the pentagon (or extensions of the sides).

    a. Show that $PV + PW + PX + PY + PZ$ does not depend on how P is chosen inside the pentagon.

    b. If $AB = 5$, find the value of $PV + PW + PX + PY + PZ$. Round to the nearest tenth.

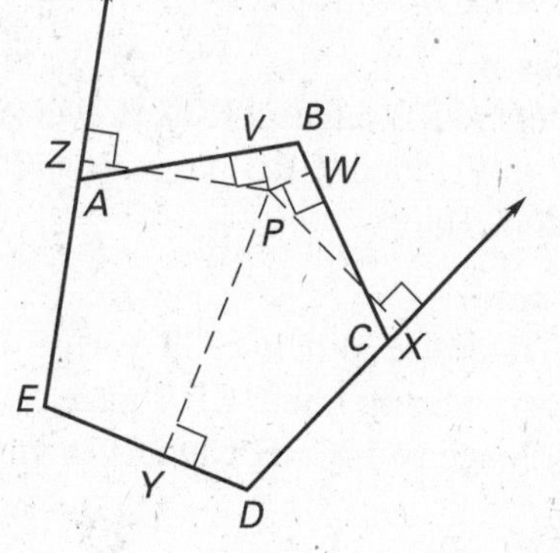

TEACHER'S NAME ____________________ CLASS ______ ROOM ______ DATE ______

Lesson Plan

2-day lesson (See *Pacing the Chapter,* TE pages 658C–658D) For use with pages 676–682

1. **Compare perimeters and areas of similar figures.**
2. **Use perimeters and areas of similar figures to solve real-life problems.**

State/Local Objectives ____

____

✓ Check the items you wish to use for this lesson.

STARTING OPTIONS

____ Homework Check: TE page 672: Answer Transparencies
____ Warm-Up or Daily Homework Quiz: TE pages 677 and 675, CRB page 40, or Transparencies

TEACHING OPTIONS

____ Motivating the Lesson: TE page 678
____ Concept Activity: SE page 676; CRB page 41 (Activity Support Master)
____ Lesson Opener (Spreadsheet): CRB page 42 or Transparencies
____ Examples: Day 1: 1–3, SE pages 677–678; Day 2: See the Extra Examples.
____ Extra Examples: Day 1 or Day 2: 1–3, TE page 678 or Transp.
____ Closure Question: TE page 678
____ Guided Practice: SE page 679 Day 1: Exs. 1–6; Day 2: See Checkpoint Exs. TE page 678

APPLY/HOMEWORK

Homework Assignment

____ Basic Day 1: 8–20 even, 23–27; Day 2: 7–21 odd, 28, 29, 34–41; Quiz 1: 1–7
____ Average Day 1: 8–20 even, 23–27; Day 2: 7–21 odd, 28, 29, 34–41; Quiz 1: 1–7
____ Advanced Day 1: 8–20 even, 23–27; Day 2: 7–21 odd, 22, 28–41; Quiz 1: 1–7

Reteaching the Lesson

____ Practice Masters: CRB pages 43–45 (Level A, Level B, Level C)
____ Reteaching with Practice: CRB pages 46–47 or Practice Workbook with Examples
____ Personal Student Tutor

Extending the Lesson

____ Applications (Interdisciplinary): CRB page 49
____ Math & History: SE page 682; CRB page 50; Internet
____ Challenge: SE page 681; CRB page 51 or Internet

ASSESSMENT OPTIONS

____ Checkpoint Exercises: Day 1 or Day 2: TE page 678 or Transp.
____ Daily Homework Quiz (11.3): TE page 681, CRB page 55, or Transparencies
____ Standardized Test Practice: SE page 681; TE page 681; STP Workbook; Transparencies
____ Quiz (11.1–11.3): SE page 682; CRB page 52

Notes ____

____

____

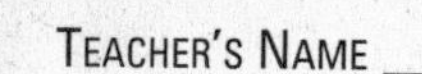

LESSON 11.3

TEACHER'S NAME ______________ CLASS ______ ROOM ______ DATE ______

Lesson Plan for Block Scheduling

1-day lesson (See *Pacing the Chapter,* TE pages 658C–658D)

For use with pages 676–682

GOALS
1. Compare perimeters and areas of similar figures.
2. Use perimeters and areas of similar figures to solve real-life problems.

State/Local Objectives ______________

CHAPTER PACING GUIDE	
Day	**Lesson**
1	11.1 (all)
2	11.2 (all)
3	**11.3 (all)**
4	11.4 (all)
5	11.5 (all); 11.6 (begin)
6	11.6 (end); Review Ch. 11
7	Assess Ch. 11; 12.1 (all)

✓ **Check the items you wish to use for this lesson.**

STARTING OPTIONS

____ Homework Check: TE page 672: Answer Transparencies
____ Warm-Up or Daily Homework Quiz: TE pages 677 and 675, CRB page 40, or Transparencies

TEACHING OPTIONS

____ Motivating the Lesson: TE page 678
____ Concept Activity: SE page 676; CRB page 41 (Activity Support Master)
____ Lesson Opener (Spreadsheet): CRB page 42 or Transparencies
____ Examples: 1–3, SE pages 677–678
____ Extra Examples: TE page 678 or Transparencies
____ Closure Question: TE page 678
____ Guided Practice Exercises: SE page 679

APPLY/HOMEWORK

Homework Assignment

____ Block Schedule: 7–21, 23–29, 34–41; Quiz 1: 1–7

Reteaching the Lesson

____ Practice Masters: CRB pages 43–45 (Level A, Level B, Level C)
____ Reteaching with Practice: CRB pages 46–47 or Practice Workbook with Examples
____ Personal Student Tutor

Extending the Lesson

____ Applications (Interdisciplinary): CRB page 49
____ Math & History: SE page 682; CRB page 50; Internet
____ Challenge: SE page 681; CRB page 51 or Internet

ASSESSMENT OPTIONS

____ Checkpoint Exercises: TE page 678 or Transparencies
____ Daily Homework Quiz (11.3): TE page 681, CRB page 55, or Transparencies
____ Standardized Test Practice: SE page 681; TE page 681; STP Workbook; Transparencies
____ Quiz (11.1–11.3): SE page 682; CRB page 52

Notes ______________

Available as
a transparency

LESSON
11.3

NAME ______________________ DATE __________

WARM-UP EXERCISES

For use before Lesson 11.3, pages 676–682

Polygons *ABCD* and *HJKL* are similar.

1. What is the ratio of the side lengths of *ABCD* to *HJKL*?
2. Find *KL*.
3. Find *AB*.
4. What is the ratio of their perimeters?

DAILY HOMEWORK QUIZ

For use after Lesson 11.2, pages 669–675

1. Find the area of the triangle.

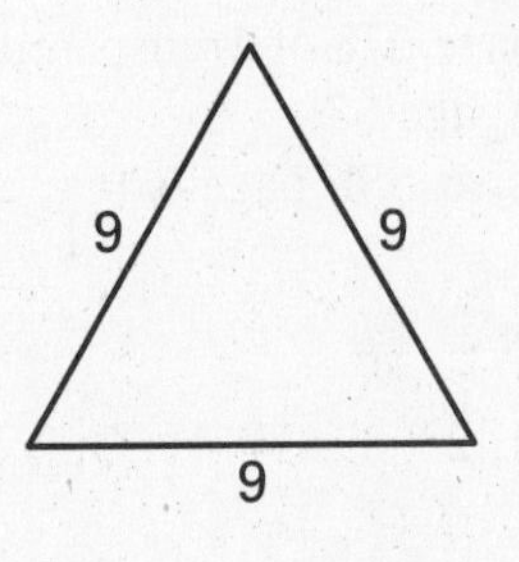

2. Find the measure of the central angle of a regular polygon with 10 sides.

3. Find the area of the inscribed regular polygon shown.

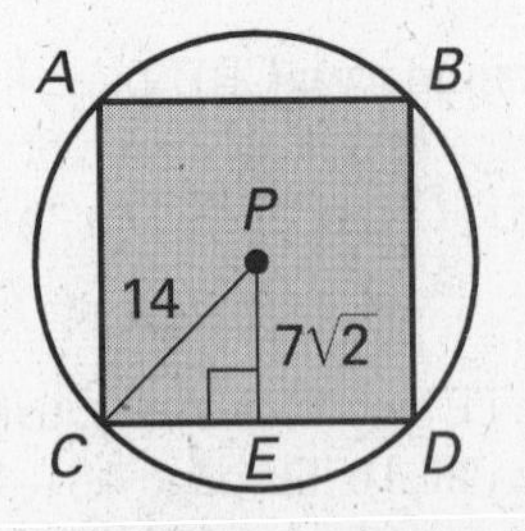

4. Find the area of a regular hexagon that has 4-inch sides.

Lesson 11.3

NAME ______________________________ DATE ____________

Activity Support Master

For use with page 676

Original Polygon	*Area*	*Similar Polygon*	*Area*
Rectangle 1		Rectangle 1	
Rectangle 2		Rectangle 2	
Rectangle 3		Rectangle 3	
Rectangle 4		Rectangle 4	
Rectangle 5		Rectangle 5	
Total		Total	

LESSON **11.3** NAME ______________________ DATE __________

Available as a transparency

Spreadsheet Lesson Opener

For use with pages 677–682

Use a spreadsheet to explore perimeters and areas of similar triangles.

1. Design a spreadsheet to calculate and compare perimeters and areas of triangles similar to a 3-4-5 triangle.

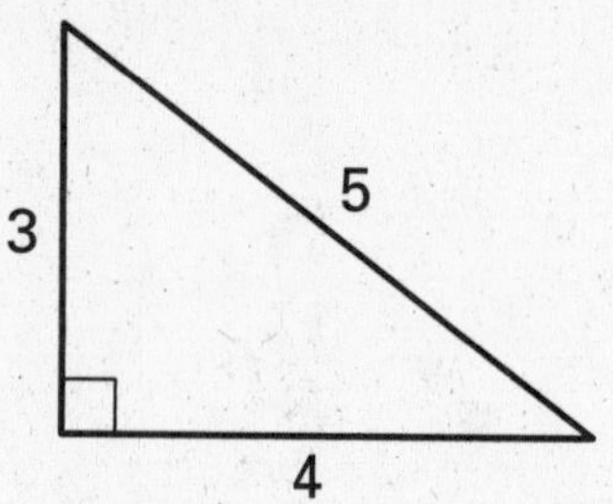

Type headings in the first row of your spreadsheet, which will have columns A through I. Name a 3-4-5 triangle as Triangle 1.

Ratios

	A	B	C	D	E	F	G	H	I
1	Triangle	a	b	c	Perimeter	Area	Ratio of sides	Ratio of perimeters	Ratio of areas
2	1	3	4	5					

Type the formulas as shown in the spreadsheet below. Describe what each formula does.

Ratios

	A	B	C	D	E
1	Triangle	a	b	c	Perimeter
2	1	3	4	5	=B2+C2+D2
3	=A2+1	=A3*3	=A3*4	=A3*5	

Ratios

F	G	H	I
Area	Ratio of sides	Ratio of perimeters	Ratio of areas
=0.5*B2*C2	=B2/3	=E2/12	=F2/6

2. Fill down all the columns to Triangle 20 by copying the formulas and analyze the numbers. Make a conjecture about the ratio of the perimeters of similar triangles and the ratio of the areas of similar triangles.

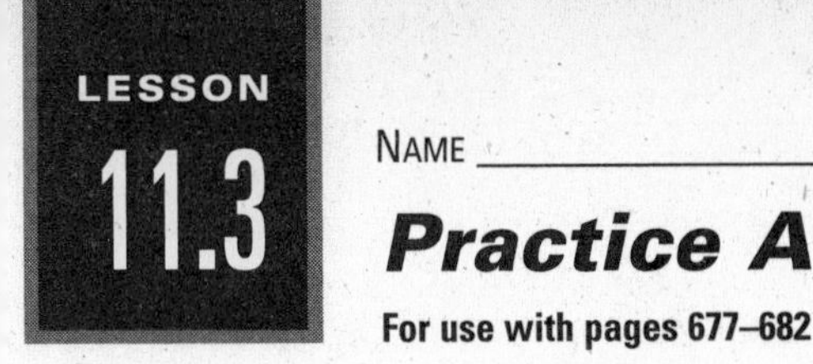

NAME ______________________ DATE __________

Practice A

For use with pages 677–682

The polygons shown are similar. Find the ratio (shaded to unshaded) of their perimeters and of their areas.

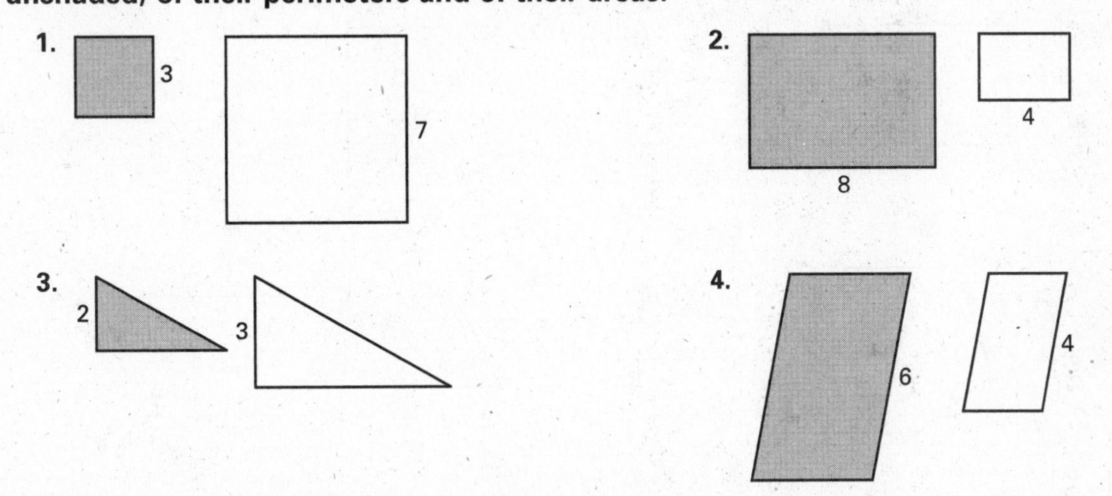

Complete the statement using *always, sometimes,* or *never.*

5. Two similar quadrilaterals ___?___ have the same perimeter.
6. Two squares with the same perimeter are ___?___ similar.
7. Two regular hexagons are ___?___ similar.
8. Two right triangles with the same area are ___?___ similar.

Solve.

9. The ratio of the lengths of corresponding sides of two similar triangles is 5:8. What is the ratio of their areas?
10. The ratio of the areas of two similar triangles is 16:9. What is the ratio of the lengths of corresponding sides?
11. A regular pentagon has an area of 48 square centimeters. Find the scale factor of this pentagon to a similar pentagon that has an area of 75 square centimeters.
12. The ratio of the lengths of corresponding sides of two similar rectangles is 3:5. The smaller rectangle has an area of 36 square centimeters. What is the area of the larger rectangle?

In Exercises 13–15, use the diagram of the room and a ruler. The scale is 1 centimeter to 1 meter.

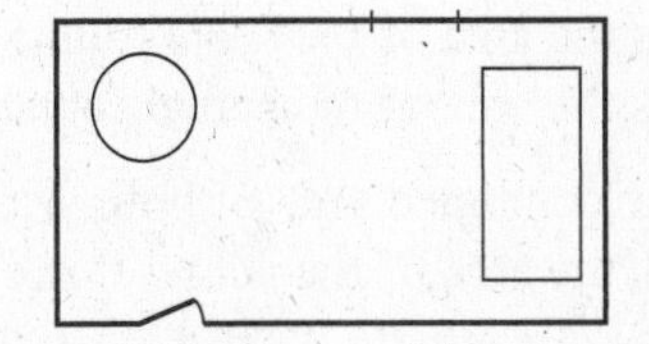

13. Use a ruler to approximate the dimensions of the room.
14. What are the dimensions of the actual room?
15. Show that the ratio of the area of the model to the area of the actual room is 1 cm^2 to 1 m^2.

Lesson 11.3

LESSON 11.3

NAME ______________________ DATE __________

Practice B

For use with pages 677–682

The polygons shown are similar. Find the ratio (shaded to unshaded) of their perimeters and of their areas.

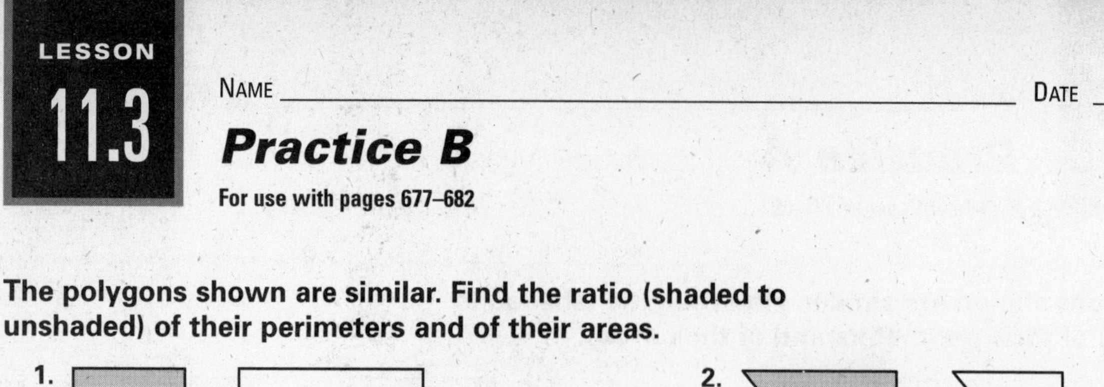

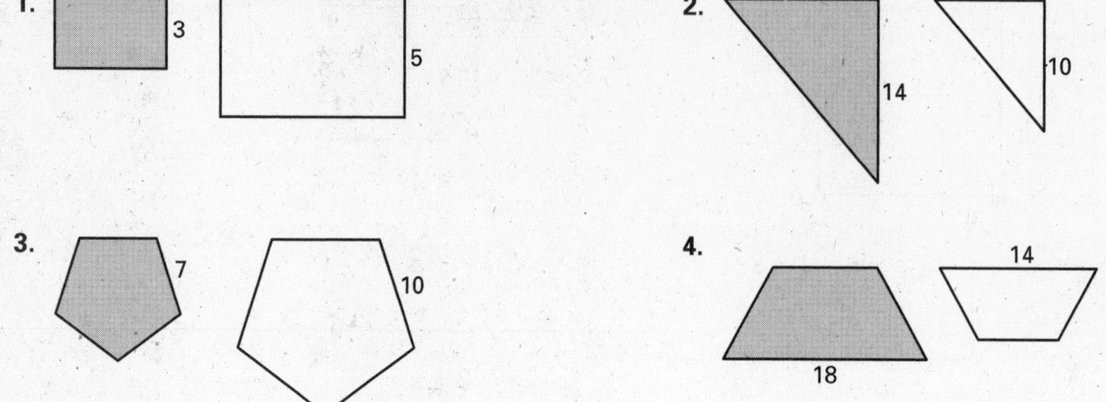

Solve.

5. The ratio of the lengths of corresponding sides of two similar polygons is 3:7. What is the ratio of their areas?

6. The ratio of the areas of two similar triangles is 32:24. What is the ratio of the lengths of corresponding sides?

7. A regular hexagon has an area of 60 square centimeters. Find the scale factor of this hexagon to a similar hexagon that has an area of 96 square centimeters.

8. The ratio of the lengths of corresponding sides of two similar triangles is 5:12. The smaller triangle has an area of 24 square centimeters. What is the area of the larger triangle?

In Exercises 9–15, use the diagram of the garden and a ruler. The scale is 1 millimeter to 0.5 meter.

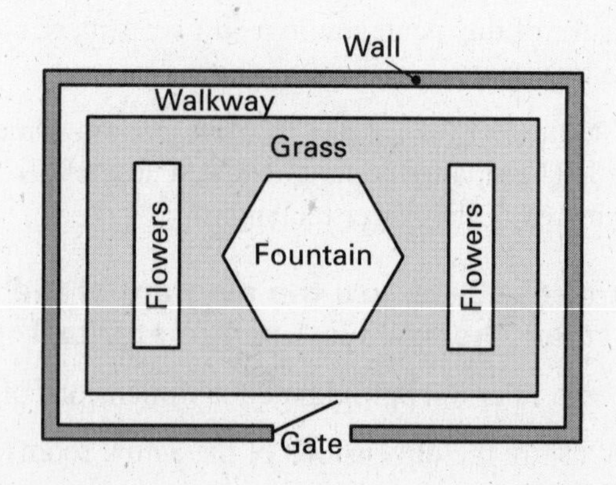

9. Use a ruler to approximate the dimensions of the scale garden including the wall.

10. Find the dimensions of the actual garden.

11. What is the area of the scale garden? What is the area of the actual garden?

12. What is the area of the scale fountain? What is the area of the actual fountain?

13. Find the combined area of both scale flower boxes. What is the area of the actual flower boxes?

14. Find the total scale area inside the walk. What is the total actual area inside the walk?

15. Find the actual area of the grass inside the garden.

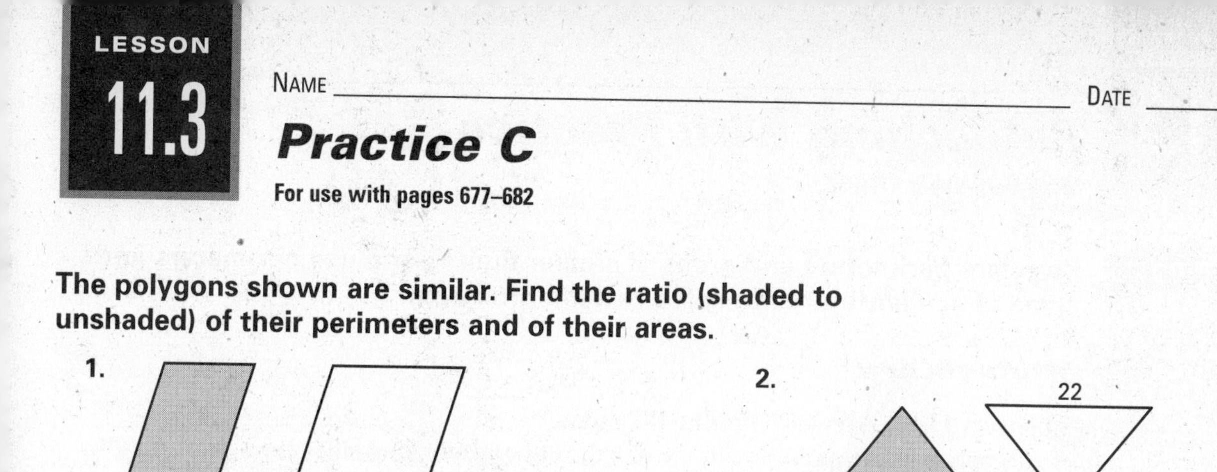

LESSON 11.3

NAME _______________ DATE _______

Practice C

For use with pages 677–682

The polygons shown are similar. Find the ratio (shaded to unshaded) of their perimeters and of their areas.

1. $\sqrt{12}$ 5

2. 28 22

3. 6 5 10

4. 15 12

Solve.

5. The perimeter of an equilateral triangle is 48 centimeters. A smaller equilateral triangle has a side length of 6 centimeters. What is the ratio of the areas of the larger triangle to the smaller triangle?

6. The ratio of the areas of two similar triangles is 84:40. What is the ratio of the lengths of corresponding sides?

7. A pentagon has an area of 128 square centimeters. A similar pentagon has an area of 180 square centimeters. What is the ratio of the perimeters of the smaller pentagon to the larger pentagon?

8. The dimensions of a rectangle are 8 centimeters by 12 centimeters. What are the dimensions of a similar rectangle with exactly double the area?

9. Find the ratio of the areas of the triangles.

24 30° 60° 8

10. Find the ratio of area I to area II.

12 I 10 II 14 18

11. ***Floor plan*** The floor plan has a scale of 1 inch to 18 feet.

a. What is the scale area of the kitchen? What is the actual area?

b. What is the scale area of the bedroom? What is the actual area?

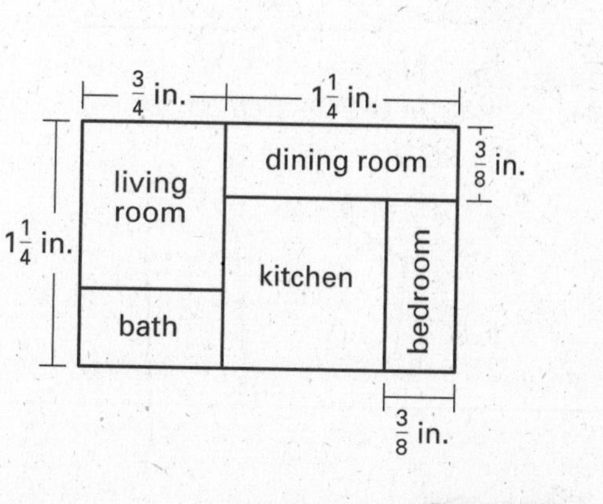

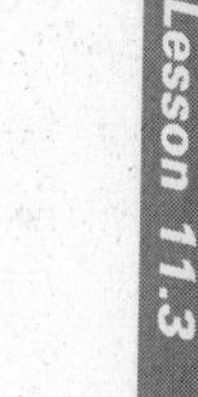

NAME ______________________ DATE ________

Reteaching with Practice

For use with pages 677–682

GOAL **Compare perimeters and areas of similar figures and use perimeters and areas of similar figures to solve real-life problems**

VOCABULARY

Theorem 11.5 Areas of Similar Polygons
If two polygons are similar with the lengths of corresponding sides in the ratio of $a:b$, then the ratio of their areas is $a^2:b^2$.

EXAMPLE 1 *Finding Ratios of Similar Polygons*

Hexagons *ABCDEF* and *LMNPQR* are similar.

a. Find the ratio of the perimeters of the hexagons.

b. Find the ratio of the areas of the hexagons.

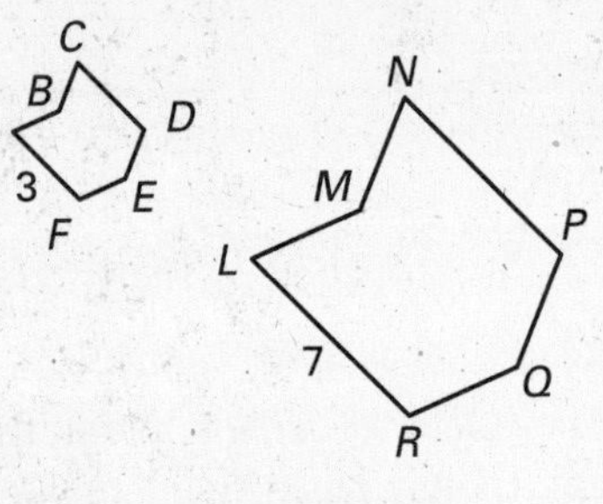

SOLUTION

The ratio of the lengths of corresponding sides in the hexagons is $\frac{3}{7}$, or 3:7.

a. The ratio of the perimeters is also 3:7. So, the perimeter of hexagon *ABCDEF* is $\frac{3}{7}$ of the perimeter of hexagon *LMNPQR*.

b. Using Theorem 11.5, the ratio of the areas is $3^2:7^2$, or 9:49. So, the area of hexagon *ABCDEF* is $\frac{9}{49}$ times the area of hexagon *LMNPQR*.

Exercises for Example 1

In Exercises 1–4, the polygons are similar. Find the ratio of their perimeters and of their areas.

1. $\triangle ABC \sim \triangle DEF$

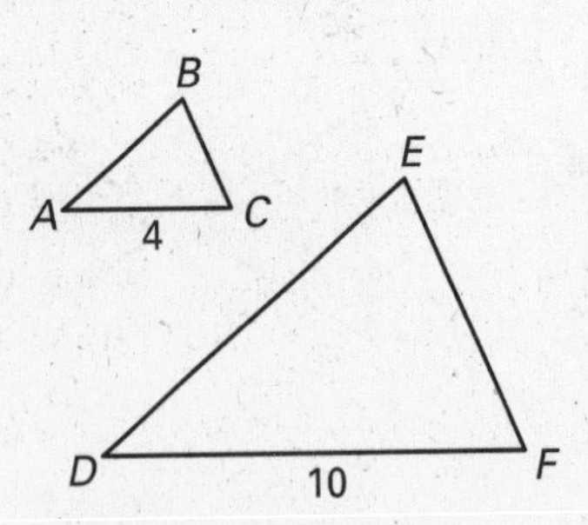

2. $ABCD \sim GHEF$

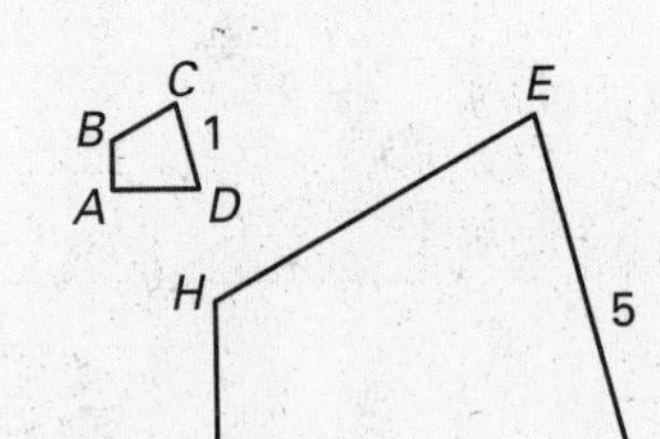

3. $JKLMN \sim PQRST$

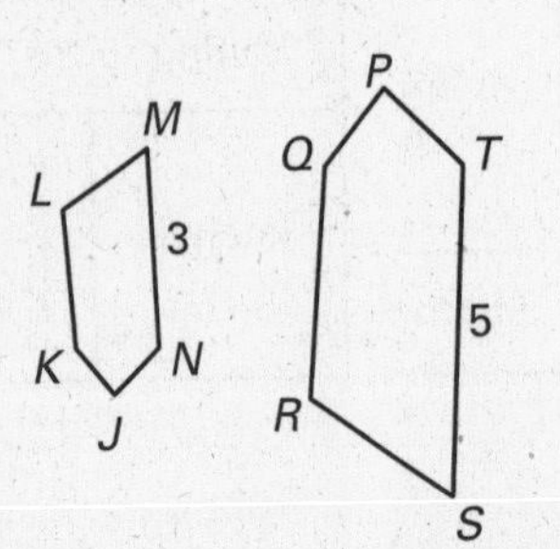

4. $\triangle LKJ \sim \triangle XYZ$

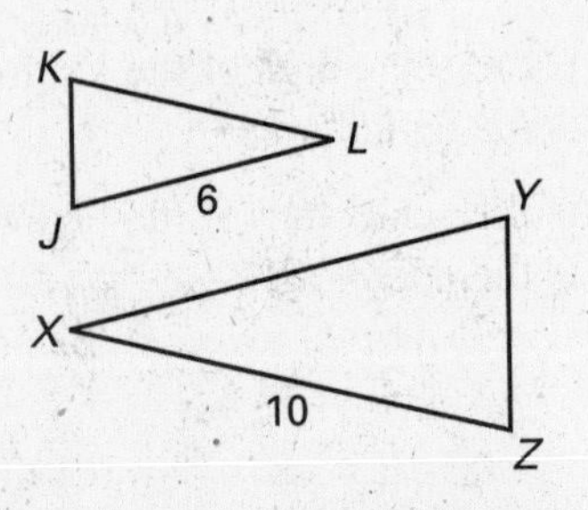

NAME ______________________________ DATE ____________

Reteaching with Practice

For use with pages 677–682

EXAMPLE 2 Using Areas of Similar Figures

A train set is designed to be $\frac{1}{12}$ actual size.

a. A billboard for the train set measures 5.4 inches by 3.2 inches. What would the dimensions of the billboard be in real life?

b. If it took Joe 6 minutes to paint the billboard for the train set, what is a rough approximation of how long it would take to paint the real-life billboard?

c. If the perimeter of the track for Joe's train were 18 feet, how long would the track be in real life?

3.2

5.4

SOLUTION

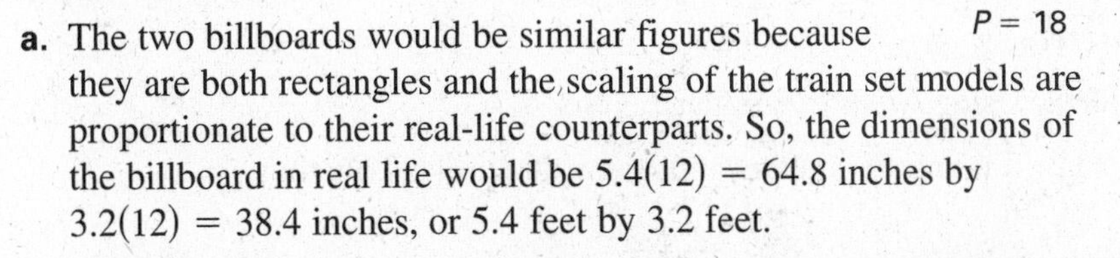

a. The two billboards would be similar figures because they are both rectangles and the scaling of the train set models are proportionate to their real-life counterparts. So, the dimensions of the billboard in real life would be $5.4(12) = 64.8$ inches by $3.2(12) = 38.4$ inches, or 5.4 feet by 3.2 feet.

b. The ratio of the areas of the rectangles is $1^2:12^2$, or 1:144. Because the amount of time it takes to paint the billboard should be a function of its area, the larger billboard should take about 144 times as long to paint. This would be 864 minutes, or 14.4 hours.

c. The ratio of perimeters of the tracks is the same as the ratio of the individual lengths, or 1:12. So, the perimeter of the real-life track would be $18(12) = 216$ feet.

Exercises for Example 2

In Exercises 5–9, refer to the situation in Example 2.

5. The real-life train that Joe modeled his train set after crosses a circular pond with radius 20.5 feet. How big of a circle should Joe draw to represent this pond in his model train set, if he is to stay consistent with the 1:12 ratio?

6. If it takes Joe's train about 7 seconds to traverse the pond in his model set, what is a good approximation of how long it would take the real-life train to cross the real pond?

7. Joe purchased some blue suede material to use as the surface of his pond. The amount of suede used to cover the pond cost about $13.10. What is a good estimate of the cost of enough suede to cover the real-life pond?

8. The area of the real pond is known to be approximately 1320.25 square feet. Use the scale ratio to approximate the area of Joe's model pond, in square feet.

9. Joe created a model of a cow that measured 11 inches long. Is this a good, proportionate representation of a real-life cow? Explain.

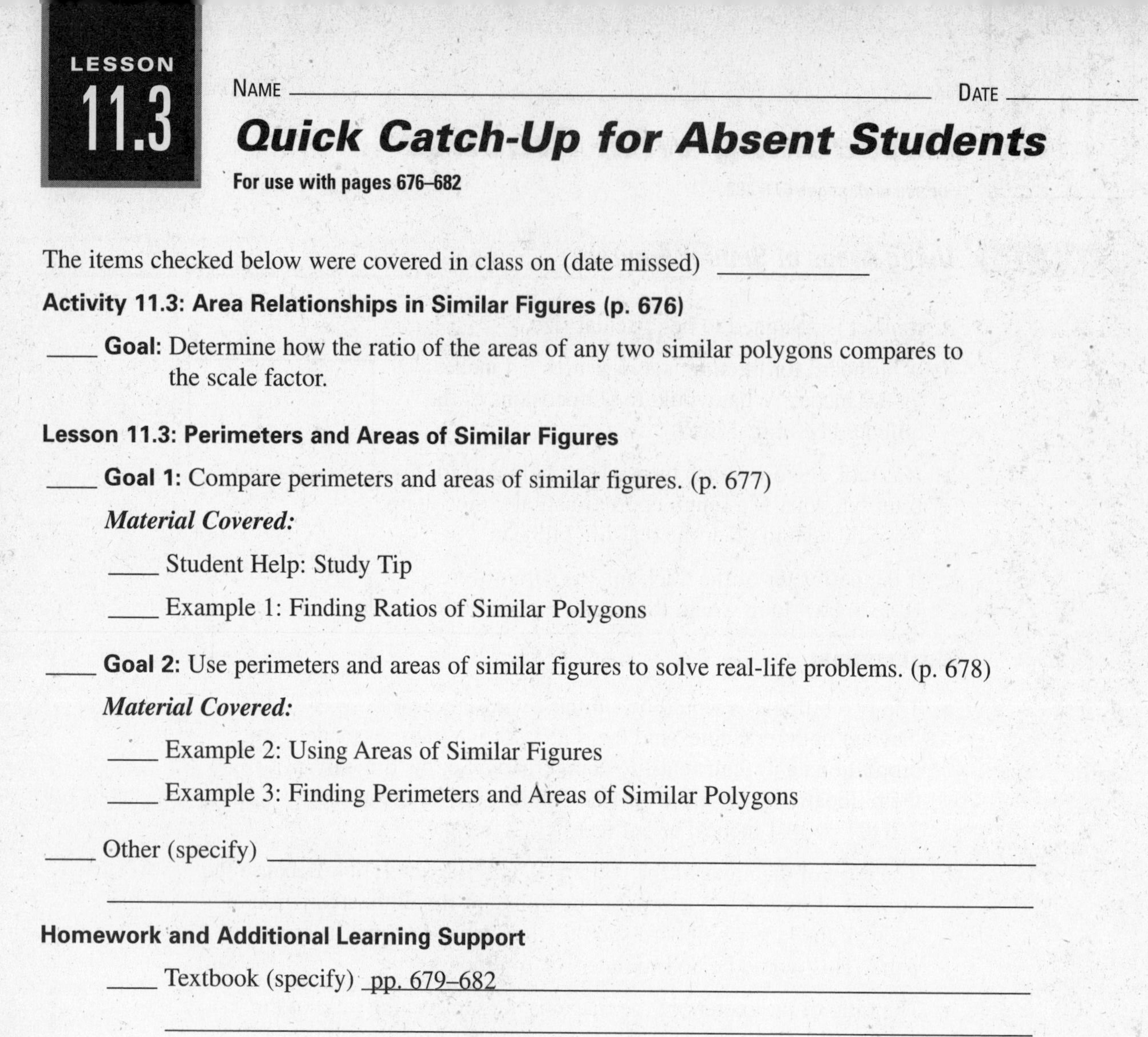

LESSON 11.3

NAME ____________________ DATE __________

Quick Catch-Up for Absent Students

For use with pages 676–682

The items checked below were covered in class on (date missed) __________

Activity 11.3: Area Relationships in Similar Figures (p. 676)

____ **Goal:** Determine how the ratio of the areas of any two similar polygons compares to the scale factor.

Lesson 11.3: Perimeters and Areas of Similar Figures

____ **Goal 1:** Compare perimeters and areas of similar figures. (p. 677)

Material Covered:

____ Student Help: Study Tip

____ Example 1: Finding Ratios of Similar Polygons

____ **Goal 2:** Use perimeters and areas of similar figures to solve real-life problems. (p. 678)

Material Covered:

____ Example 2: Using Areas of Similar Figures

____ Example 3: Finding Perimeters and Areas of Similar Polygons

____ Other (specify) ____________________

____

Homework and Additional Learning Support

____ Textbook (specify) pp. 679–682

____

____ *Reteaching with Practice* worksheet (specify exercises) __________

____ *Personal Student Tutor* for Lesson 11.3

NAME ______________________________ DATE ____________

Interdisciplinary Application

For use with pages 677–682

Photography

CAMERAS A French physicist, Joseph Nicephore Niepce, using a metal plate coated with a light-sensitive chemical, made the world's first photograph in 1826. Niepce's process was modified by several people up until the late 1800s, with the introduction of the Kodak box camera by George Eastman. Eastman's invention led to a tremendous rise of amateur photographers. Eastman also began the process of taking pictures on a roll of film and sending the film (in the camera) to a processing plant. Since this time, the photographic process has evolved and become even more convenient. Today, taking a photograph is as easy as aiming the camera and pushing a button.

One of the most important gifts of photography is that it can present us with visual images of places where one might never be able to go, such as photos taken by satellites deep in space, photos of people in foreign countries, and even photos from inside our own bodies.

In Exercises 1–3, use the following information.

An instrument, called an enlarger, is used to produce photographs that are larger than their negatives. The enlarger projects an image onto photographic paper. The size of this projected image depends on the distance between the negative image placed in the enlarger and the photographic paper.

Suppose you are enlarging photos for your school yearbook. Some examples of enlargement sizes are 5-inch by 7-inch, 8-inch by 10-inch, 8-inch by 12-inch, and poster-size photos of 16-inch by 20-inch, and 20-inch by 32-inch.

1. If you have an 8-inch by 10-inch photo that was enlarged from a 4-inch by 5-inch print, what is the ratio of the lengths of the sides? What is the ratio of the areas?
2. Find the ratio of the areas of a 5-inch by 8-inch photo and a 20-inch by 32-inch poster. Is the ratio of the lengths of the sides the same as the ratio of the perimeters?
3. You want to make a poster of an 8-inch by 10-inch photo that would be 16-inch by 20-inch. If an 8-inch by 10-inch piece of photographic paper costs $0.48, what would you expect to pay for a 16-inch by 20-inch sheet? a 20-inch by 32-inch sheet?

NAME ______________________ DATE ________

Math and History Application

For use with page 682

HISTORY One of the greatest mathematicians of ancient Greece was Archimedes (287 B.C.–212 B.C.). Born in Syracuse, a Greek settlement on the island of Sicily, Archimedes' inventions were extremely useful in defending Syracuse from Roman invasion in 212 B.C. These inventions included catapults, poles which extended from the city walls and dropped heavy weights on approaching ships, and cranes which lifted ships up by one end and dropped them into the water.

MATH One of Archimedes' most important contributions to mathematics was his approximation of π. He observed that, if regular polygons are inscribed and circumscribed about a circle, then

$$\begin{array}{c}\text{perimeter}\\\text{of inscribed}\\\text{polygon}\end{array} < \begin{array}{c}\text{circumference}\\\text{of circle}\end{array} < \begin{array}{c}\text{perimeter}\\\text{of circumscribed}\\\text{polygon}\end{array}.$$

In particular, Archimedes considered a circle of diameter 1, with inscribed and circumscribed hexagons as shown below.

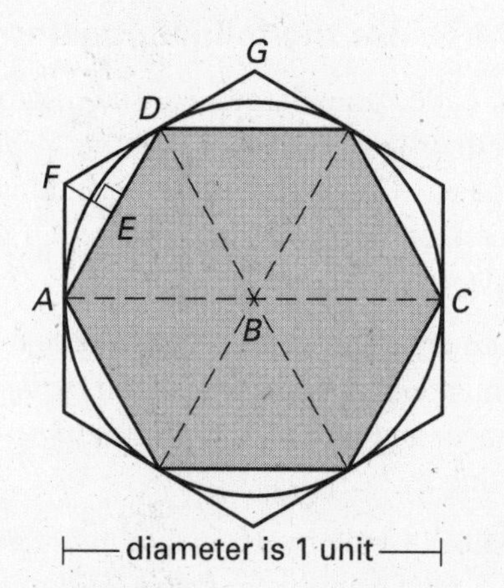

1. Show that the perimeter of the inscribed hexagon is 3.
2. Recall that, in a 30° - 60° - 90° triangle, the sides are of length x, $\sqrt{3}x$, and $2x$. Use this fact to show that the length of a side of the circumscribed hexagon is $\frac{\sqrt{3}}{3}$.
3. Show that the perimeter of the circumscribed hexagon is $2\sqrt{3}$.
4. Show that Exercise 3 leads to the approximation $3 < \pi < 2\sqrt{3}$.

NAME ______________________ DATE __________

Challenge: Skills and Applications

For use with pages 677–682

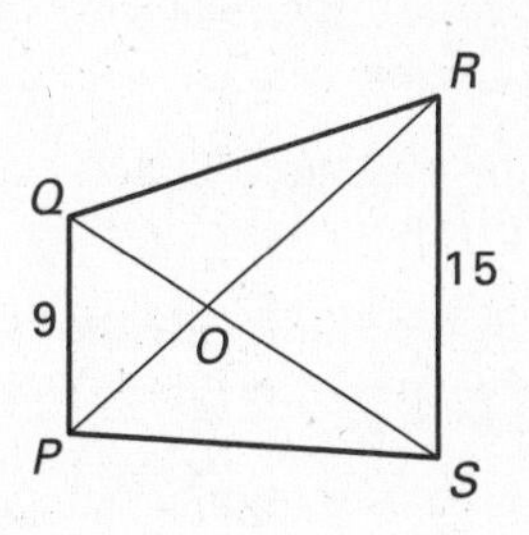

1. Refer to the diagram. $PQRS$ is a trapezoid with bases $\overline{PQ}$, of length 9, and $\overline{RS}$, of length 15. The area of $\triangle OPQ$ is 27.
   a. Find the area of $\triangle ORS$.
   b. Find the height of $\triangle OPQ$.
   c. Find the height of $\triangle ORS$.
   d. Find the area of trapezoid $PQRS$.

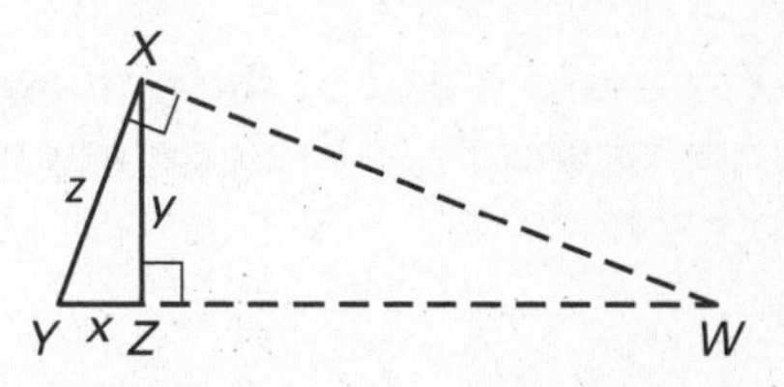

2. Areas of similar triangles can be used to prove the Pythagorean Theorem. Let $\triangle XYZ$ be a right triangle with side lengths x, y, and z, as shown. Extend $\overline{YZ}$ to W so that $\overline{WX} \perp \overline{XY}$. As you know, $\triangle XYZ \sim \triangle WXZ \sim \triangle WYX$.
   a. Find the ratios $\dfrac{\text{Side length of } \triangle WXZ}{\text{Side length of } \triangle XYZ}$ and $\dfrac{\text{Side length of } \triangle WYX}{\text{Side length of } \triangle XYZ}$ in terms of x, y, and z.
   b. Let A be the area of $\triangle XYZ$. Find expressions for the area of $\triangle WXZ$ and the area of $\triangle WYX$ in terms of A, x, y, and z.
   c. Observe that area of $\triangle XYZ$ + area of $\triangle WXZ$ = area of $\triangle WYX$. Explain how to use this fact to derive the Pythagorean Theorem.

In Exercises 3–6, use the information about a pair of similar polygons to find all possible values of *x*.

3. The area of $\triangle ABC$ is 5. The perimeter of $\triangle ABC$ is 2.
   The area of $\triangle DEF$ is $x^2 + 9$. The perimeter of $\triangle DEF$ is x.
4. The area of $HIJK$ is $x + 4$. The perimeter of $HIJK$ is $\frac{x}{2}$.
   The area of $STUV$ is 5. The perimeter of $STUV$ is $\sqrt{x}$.
5. The area of $\triangle PQR$ is $2x - 11$. The perimeter of $\triangle PQR$ is $x - 7$.
   The area of $\triangle WXY$ is $2x + 5$. The perimeter of $\triangle WXY$ is $x - 5$.
6. The area of $EFGH$ is $2x + 9$. The perimeter of $EFGH$ is $x + 3$.
   The area of $JKLM$ is $2x - 6$. The perimeter of $JKLM$ is $x - 1$.

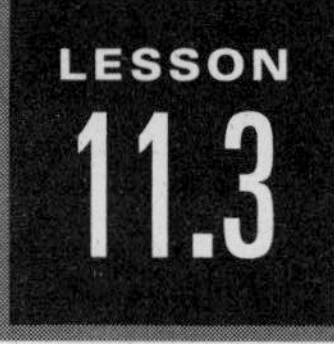

NAME ______________________ DATE __________

Quiz 1

For use after Lessons 11.1–11.3

1. Find the sum of the measures of the interior angles of a convex 24-gon. *(Lesson 11.1)*

2. What is the measure of each exterior angle of a regular 24-gon? *(Lesson 11.1)*

3. Find the area of an equilateral triangle with a side length of 20 inches. *(Lesson 11.2)*

4. Find the area of a regular octagon with an apothem of 15 centimeters. *(Lesson 11.2)*

In Exercises 5 and 6, the polygons are similar. Find the ratio (large to small) of their perimeters and of their areas. *(Lesson 11.3)*

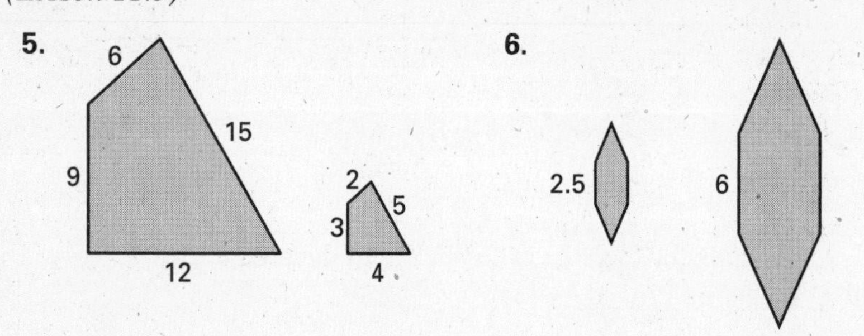

7. ***Carpet*** You just carpeted a 12 foot by 15 foot room for $1080. About how much would you expect to pay for the same carpet in another room that is 8 foot by 6 foot? *(Lesson 11.3)*

Answers

1. ______
2. ______
3. ______
4. ______
5. ______
6. ______
7. ______

Lesson 11.3

LESSON 11.4

TEACHER'S NAME ______________________ CLASS ______ ROOM ______ DATE ______

Lesson Plan

2-day lesson (See *Pacing the Chapter,* TE pages 658C–658D) **For use with pages 683–690**

GOALS

1. Find the circumference of a circle and the length of a circular arc.
2. Use circumference and arc length to solve real-life problems.

State/Local Objectives ____

____

✓ Check the items you wish to use for this lesson.

STARTING OPTIONS

____ Homework Check: TE page 679: Answer Transparencies
____ Warm-Up or Daily Homework Quiz: TE pages 683 and 681, CRB page 55, or Transparencies

TEACHING OPTIONS

____ Lesson Opener (Application): CRB page 56 or Transparencies
____ Technology Activity with Keystrokes: CRB pages 57–58
____ Examples: Day 1: 1–3, SE pages 683–684; Day 2: 4–5, SE page 685
____ Extra Examples: Day 1: TE page 684 or Transp.; Day 2: TE page 685 or Transp.; Internet
____ Technology Activity: SE page 690
____ Closure Question: TE page 685
____ Guided Practice: SE page 686 Day 1: Exs. 1–11; Day 2: Exs. 12–14

APPLY/HOMEWORK

Homework Assignment

____ Basic Day 1: 15–31; Day 2: 32–41, 44, 46–49, 52–62
____ Average Day 1: 15–31; Day 2: 32–49, 52–62
____ Advanced Day 1: 15–31; Day 2: 32–62

Reteaching the Lesson

____ Practice Masters: CRB pages 59–61 (Level A, Level B, Level C)
____ Reteaching with Practice: CRB pages 62–63 or Practice Workbook with Examples
____ Personal Student Tutor

Extending the Lesson

____ Applications (Real-Life): CRB page 65
____ Challenge: SE page 689; CRB page 66 or Internet

ASSESSMENT OPTIONS

____ Checkpoint Exercises: Day 1: TE page 684 or Transp.; Day 2: TE page 685 or Transp.
____ Daily Homework Quiz (11.4): TE page 689, CRB page 69, or Transparencies
____ Standardized Test Practice: SE page 689; TE page 689; STP Workbook; Transparencies

Notes ____

____

____

LESSON 11.4

TEACHER'S NAME ______________ CLASS ______ ROOM ______ DATE ______

Lesson Plan for Block Scheduling

1-day lesson (See *Pacing the Chapter,* TE pages 658C–658D) For use with pages 683–690

GOALS

1. Find the circumference of a circle and the length of a circular arc.
2. Use circumference and arc length to solve real-life problems.

State/Local Objectives ______________

CHAPTER PACING GUIDE	
Day	**Lesson**
1	11.1 (all)
2	11.2 (all)
3	11.3 (all)
4	**11.4 (all)**
5	11.5 (all); 11.6 (begin)
6	11.6 (end); Review Ch. 11
7	Assess Ch. 11; 12.1 (all)

✓ **Check the items you wish to use for this lesson.**

STARTING OPTIONS

____ Homework Check: TE page 679: Answer Transparencies
____ Warm-Up or Daily Homework Quiz: TE pages 683 and 681, CRB page 55, or Transparencies

TEACHING OPTIONS

____ Lesson Opener (Application): CRB page 56 or Transparencies
____ Technology Activity with Keystrokes: CRB pages 57–58
____ Examples 1–5: SE pages 683–685
____ Extra Examples: TE pages 684–685 or Transparencies; Internet
____ Technology Activity: SE page 690
____ Closure Question: TE page 685
____ Guided Practice Exercises: SE page 686

APPLY/HOMEWORK

Homework Assignment

____ Block Schedule: 15–48, 51–61

Reteaching the Lesson

____ Practice Masters: CRB pages 59–61 (Level A, Level B, Level C)
____ Reteaching with Practice: CRB pages 62–63 or Practice Workbook with Examples
____ Personal Student Tutor

Extending the Lesson

____ Applications (Real-Life): CRB page 65
____ Challenge: SE page 689; CRB page 66 or Internet

ASSESSMENT OPTIONS

____ Checkpoint Exercises: TE pages 684–685 or Transparencies
____ Daily Homework Quiz (11.4): TE page 689, CRB page 69, or Transparencies
____ Standardized Test Practice: SE page 689; TE page 689; STP Workbook; Transparencies

Notes ______________

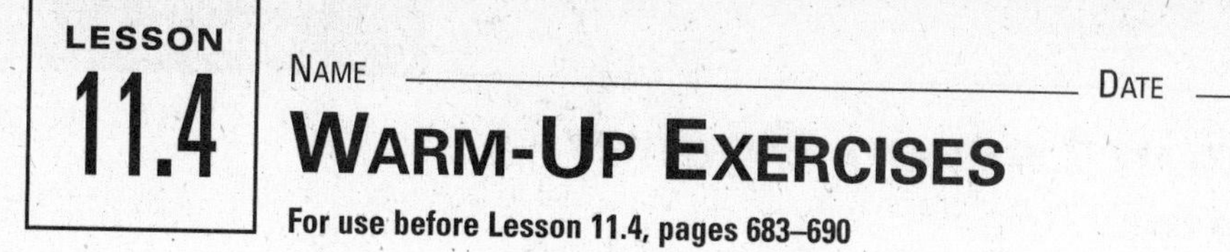

LESSON **11.4**

NAME ______________________ DATE __________

Available as a transparency

WARM-UP EXERCISES

For use before Lesson 11.4, pages 683–690

Find the measure of each arc.

1. $\overset{\frown}{JK}$

2. $\overset{\frown}{JL}$

3. $\overset{\frown}{JML}$

4. $\overset{\frown}{JMK}$

DAILY HOMEWORK QUIZ

For use after Lesson 11.3, pages 676–682

1. The polygons are similar. Find the ratio (grey to white) of their perimeters and their areas.

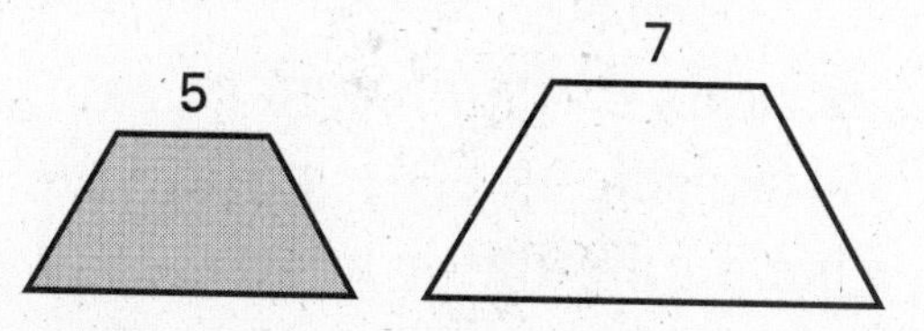

2. Determine whether $\triangle ACD$ is similar to $\triangle AEB$. If it is, find the area of $\triangle AEB$.

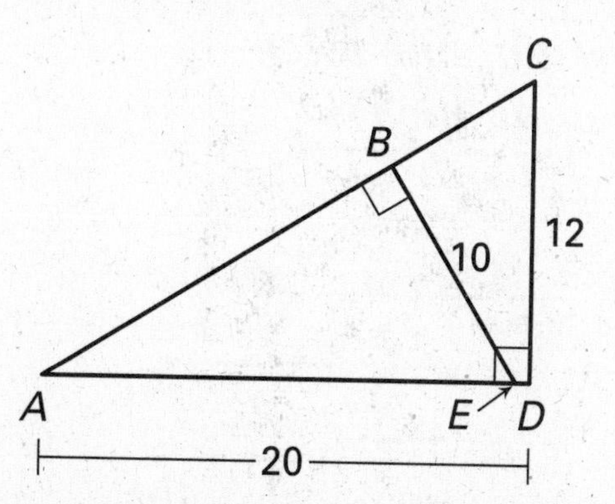

3. The price of a small tablecloth (4 ft by 5 ft) is $45 and the price of a large tablecloth (96 in. by 120 in.) is $95. Compare the costs. Is the large tablecloth a good buy? Explain.

Lesson 11.4

NAME _______________ DATE _______

Available as a transparency

Application Lesson Opener

For use with pages 683–689

YOU WILL NEED: • graph paper • ruler

An Italian restaurant offers pizzas in the four sizes shown. The table gives the number of equal pieces each pizza is cut into when it is served.

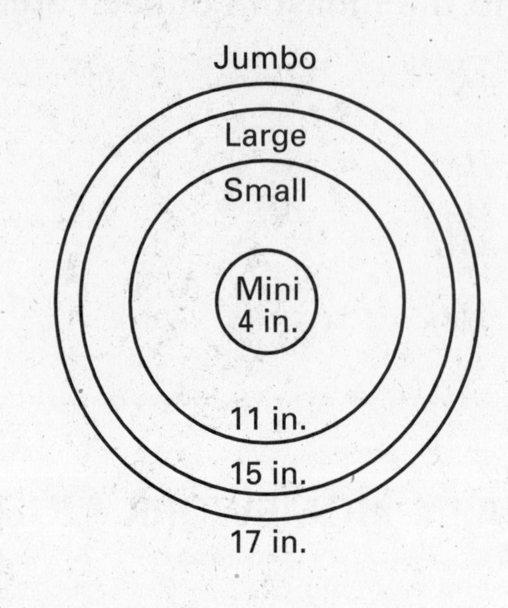

Diameter of pizza (in inches)	*Number of equal pieces*
4	4
11	6
15	8
17	12

1. Draw each size pizza on graph paper, with 1 unit = 1 inch. Show cut lines for all the pieces of pizza.

2. The distance around the edge of a pizza is its *circumference*. The formula for the circumference C of a circle is $C = \pi d$, where d is the diameter of the circle. Use a calculator to find the length of the outer crust for one piece of pizza of each size. Give your answer to the nearest sixteenth of an inch. After you complete the table, examine these lengths on your drawings.

Diameter of pizza (in inches)	*Length of outer crust of one piece (in sixteenths of an inch)*
4	
11	
15	
17	

3. Compare lengths of outer crusts by drawing a segment of that length for each. Explain why, in terms of outer crust length, a larger pizza is cut into more pieces of pizza than a smaller pizza.

NAME ______________________________ DATE __________

Technology Activity Keystrokes

For use with page 690

EXCEL

Select cell A1.

Number of sides **TAB** Perimeter **ENTER**

Select cell A2.

n **SPACE BAR** **TAB** 2*n*sin(180/n) **ENTER**

Select cell A3.

3 **TAB** = 2*A3*sin(180/A3) **ENTER**

[*Note:* If your spreadsheet uses radian measure, use "pi()" instead of 180 in the perimeter formula above. So, cell B3 should be "= 2*A3*sin(pi()/A3)"]

Select cell A4.

= A3 + 1 **ENTER**

Select cell A4. From the **Edit** menu, choose **Copy**.

Select cells A5–A24. From the **Edit** menu, choose **Paste**.

Select cell B3. From the **Edit** menu, choose **Copy**.

Select cells B4–B24. From the **Edit** menu, choose **Paste**.

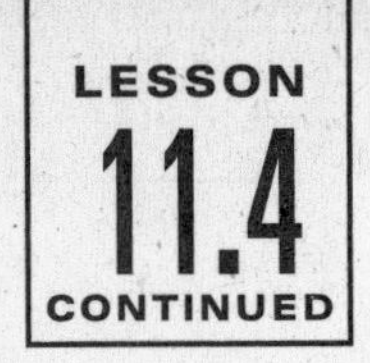

NAME _______________________________________ DATE ____________

Technology Activity Keystrokes

For use with page 690

TI-82

STAT 1

Enter the number of sides, 3 through 24, in L1.

3 ENTER 4 ENTER 5 ENTER
6 ENTER 7 ENTER 8 ENTER
9 ENTER 10 ENTER 11 ENTER
12 ENTER 13 ENTER 14 ENTER
15 ENTER 16 ENTER 17 ENTER
18 ENTER 19 ENTER 20 ENTER
21 ENTER 22 ENTER 23 ENTER
24 ENTER

2nd [QUIT]

Use a formula to find the perimeters. Store the perimeters in L2.

2 × 2nd [L1] × SIN (180 ÷
2nd [L1]) STO▸ 2nd [L2]

TI-83

STAT 1

Enter the number of sides, 3 through 24, in L1.

3 ENTER 4 ENTER 5 ENTER
6 ENTER 7 ENTER 8 ENTER
9 ENTER 10 ENTER 11 ENTER
12 ENTER 13 ENTER 14 ENTER
15 ENTER 16 ENTER 17 ENTER
18 ENTER 19 ENTER 20 ENTER
21 ENTER 22 ENTER 23 ENTER
24 ENTER

2nd [QUIT]

Use a formula to find the perimeters. Store the perimeters in L2.

2 × 2nd [L1] × SIN (180 ÷
2nd [L1]) STO▸ 2nd [L2]

LESSON 11.4

NAME ______________________ DATE ________

Practice C

For use with pages 683–689

Find the indicated measure.

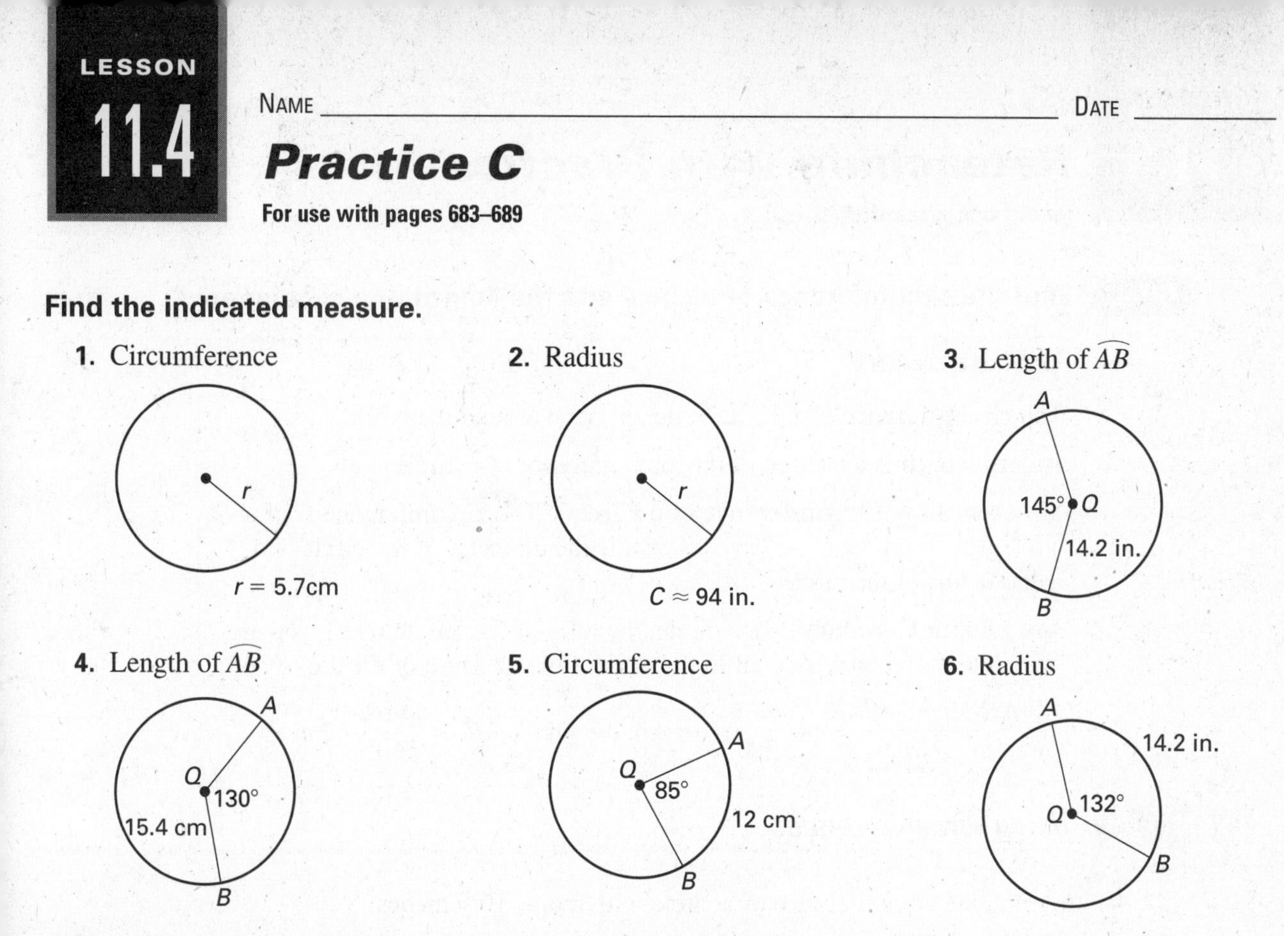

1. Circumference

2. Radius

3. Length of $\widehat{AB}$

4. Length of $\widehat{AB}$

5. Circumference

6. Radius

Each region is bounded by circular arcs or line segments. Find the perimeter of the region.

7.

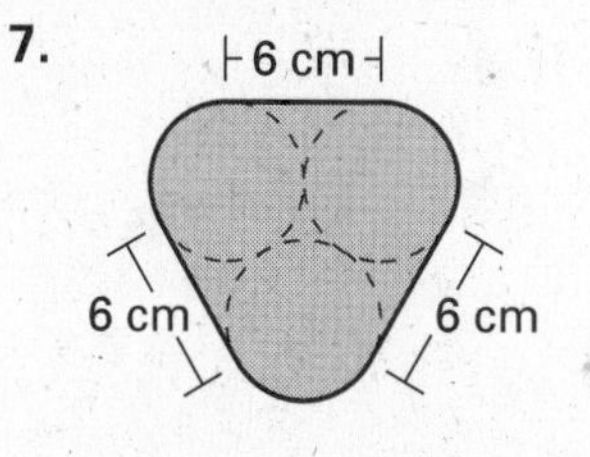

8.

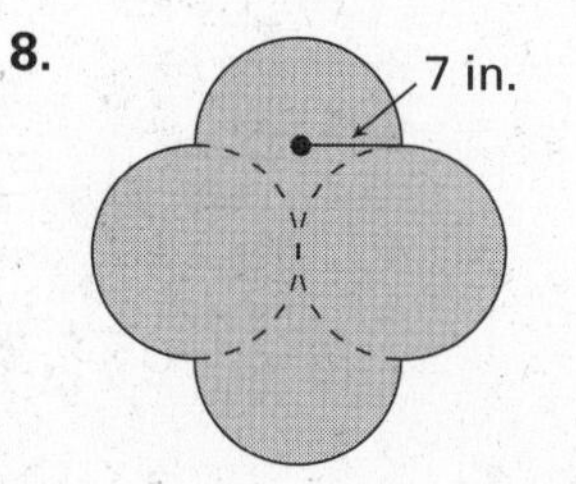

9.

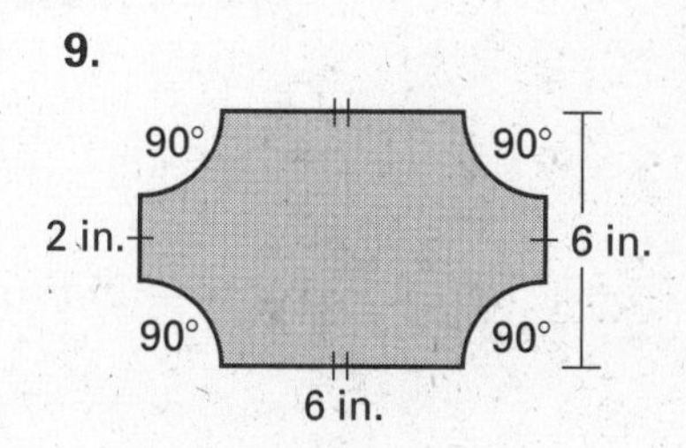

10. ***Thread*** A spool of thread contains 150 revolutions of thread. The diameter of the spool is 3 centimeters. Find the length of the thread to the nearest centimeter.

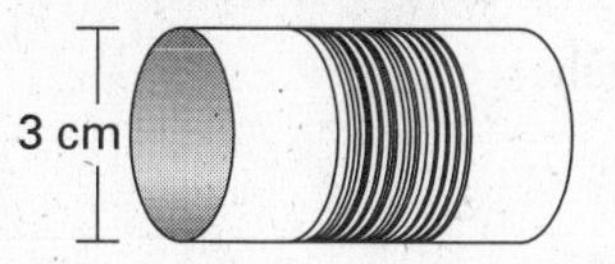

11. ***Go-Cart Track*** Find the distance around the track on the inside lane and on the outside lane.

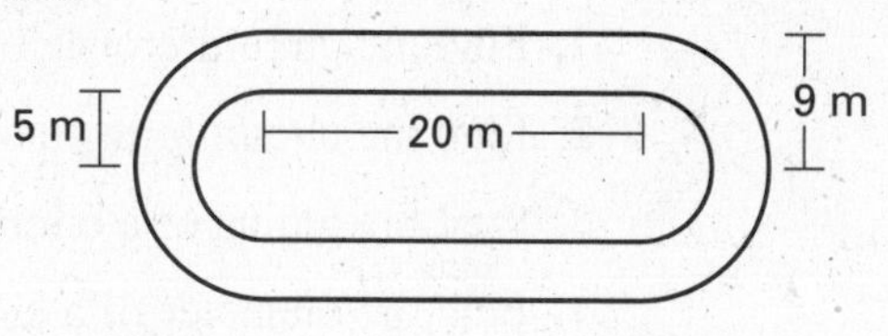

12. ***Pendulum*** Find the distance traveled in one back-and-forth swing by the weight of a 16 inch pendulum that swings through a 70° angle.

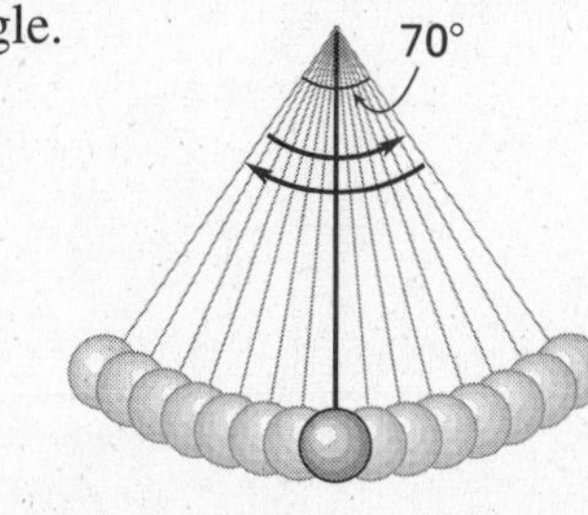

13. ***Turntable*** Two belt-driven gears for a turntable are shown. What is the total length of the belt?

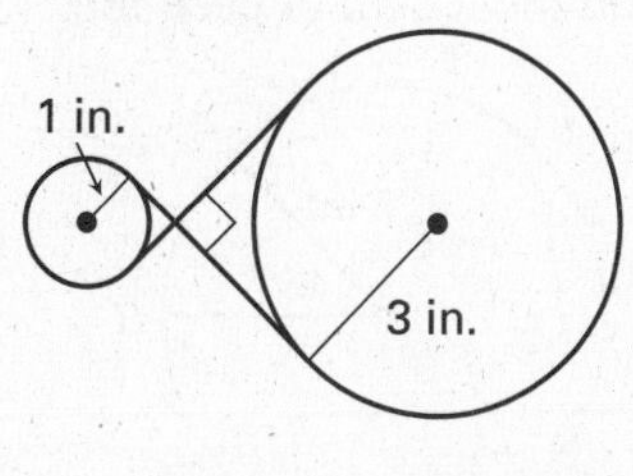

Lesson 11.4

NAME _______________ DATE _______

Reteaching with Practice

For use with pages 683–689

GOAL Find the circumference of a circle and the length of a circular arc

VOCABULARY

The **circumference** of a circle is the distance around the circle.

An **arc length** is a portion of the circumference of a circle.

Theorem 11.6 Circumference of a Circle The circumference C of a circle is $C = \pi d$ or $C = 2\pi r$, where d is the diameter of the circle and r is the radius of the circle.

Arc Length Corollary In a circle, the ratio of the length of a given arc to the circumference is equal to the ratio of the measure of the arc to 360°.

$$\frac{\text{Arc length of } \widehat{AB}}{2\pi r} = \frac{m\widehat{AB}}{360°}, \text{ or Arc length of } \widehat{AB} = \frac{m\widehat{AB}}{360°} \cdot 2\pi r$$

EXAMPLE 1 Using Circumferences

a. Find the circumference of a circle with radius 10.5 inches.

b. Find the radius of a circle with circumference 25 feet.

SOLUTION

a. $C = 2\pi r$

$C = 2 \cdot \pi \cdot (10.5)$

$C = 21\pi$

$C \approx 65.97$ inches

b. $C = 2\pi r$

$25 = 2\pi r$

$\frac{25}{2\pi} = r$

$r \approx 3.98$ feet

Exercises for Example 1

In Exercises 1–4, find the indicated measure.

1. Find the circumference of a circle with radius 17 centimeters.
2. Find the circumference of a circle with diameter 14 inches.
3. Find the radius of a circle with circumference 14 yards.
4. Find the diameter of a circle with circumference 12 feet.

EXAMPLE 2 Finding Arc Lengths

Find the length of each arc.

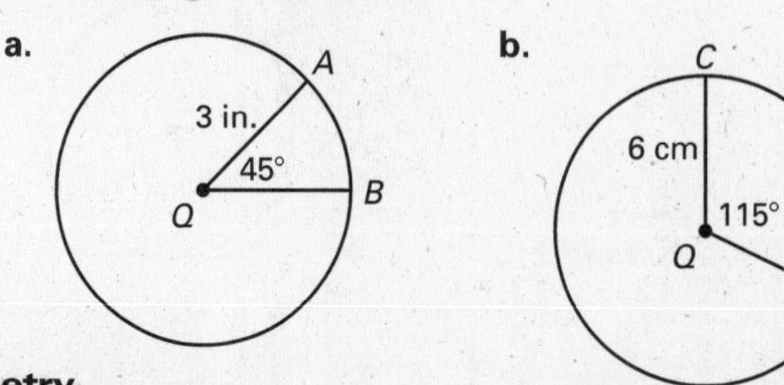

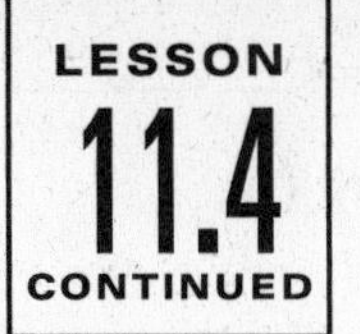

Reteaching with Practice

For use with pages 683–689

SOLUTION

a. Arc length of $\widehat{AB} = \dfrac{45^\circ}{360^\circ} \cdot 2\pi(3) \approx 2.36$ inches

b. Arc length of $\widehat{CD} = \dfrac{115^\circ}{360^\circ} \cdot 2\pi(6) \approx 12.04$ centimeters

Exercises for Example 2

In Exercises 5–7, find the length of each arc.

5.

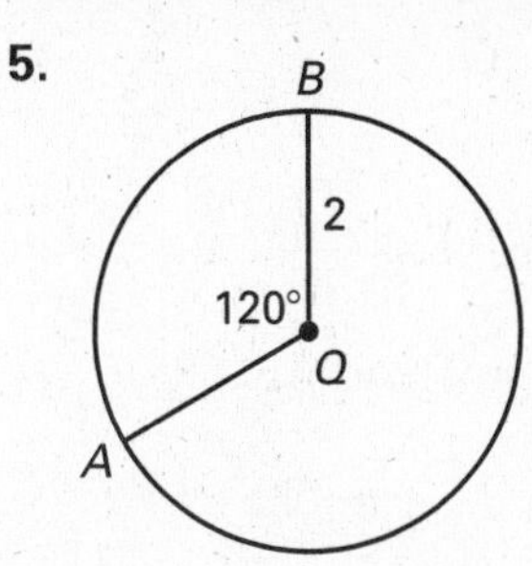

6.

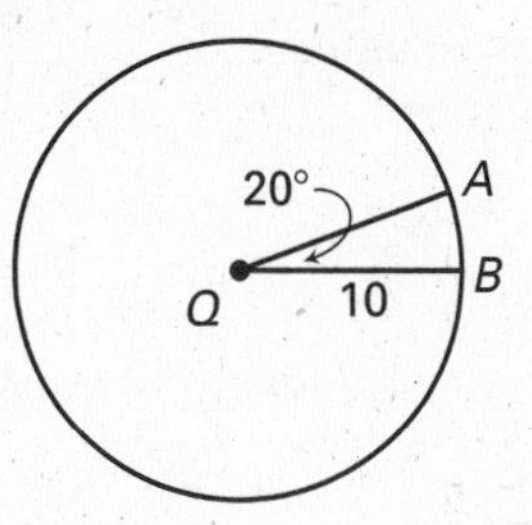

7.

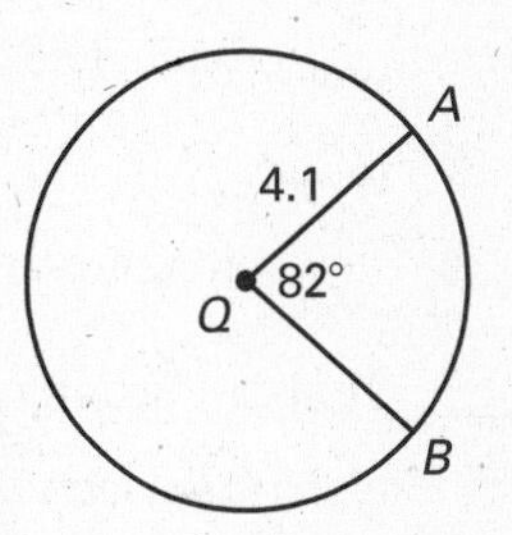

EXAMPLE 3 *Using Arc Lengths*

Find the circumference of the circle.

P
4.5
72°
R
Q

SOLUTION

$$\frac{\text{Arc length of } \widehat{PQ}}{2\pi r} = \frac{m\widehat{PQ}}{360^\circ}$$

Now substitute 4.5 for the arc length of $\widehat{PQ}$, 72° for $m\widehat{PQ}$, and C for $2\pi r$.

So, $\dfrac{4.5}{C} = \dfrac{72^\circ}{360^\circ}$, or $\dfrac{4.5}{C} = 0.2$. So, $C = \dfrac{4.5}{0.2} = 22.5$.

Exercises for Example 3

Find the indicated measure.

8. Circumference

A
Q
135°
9.8
B

9. Radius

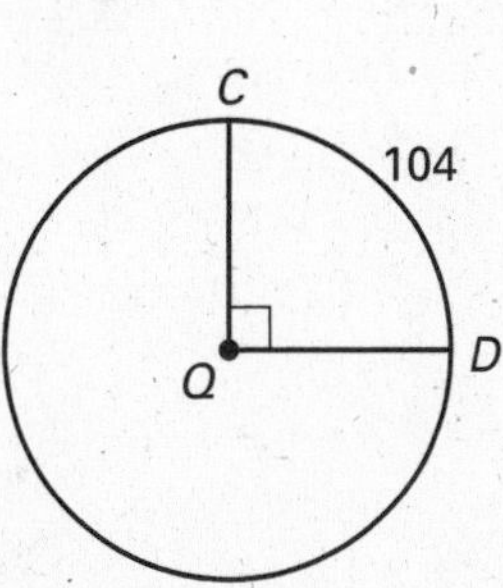

10. $m\widehat{PQ}$

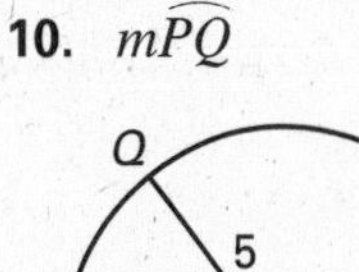

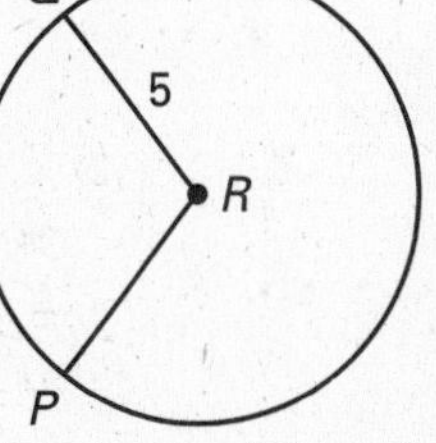

Lesson 11.4

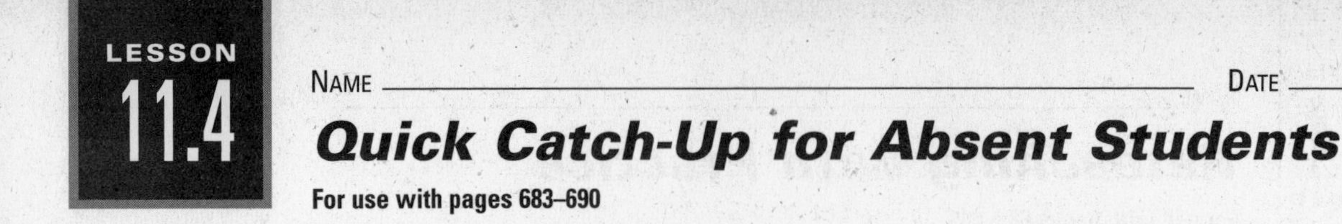

NAME ______________________________ DATE __________

Quick Catch-Up for Absent Students

For use with pages 683–690

The items checked below were covered in class on (date missed) __________

Lesson 11.4: Circumference and Arc Length

____ **Goal 1:** Find the circumference of a circle and the length of a circular arc. (pp. 683–684)

Material Covered:

____ Example 1: Using Circumference

____ Student Help: Study Tip

____ Example 2: Finding Arc Lengths

____ Example 3: Using Arc Lengths

Vocabulary:

circumference, p. 683 arc length, p. 683

____ **Goal 2:** Use circumference and arc length to solve real-life problems. (p. 685)

Material Covered:

____ Example 4: Comparing Arc Lengths

____ Example 5: Find Arc Length

Activity 11.4: Perimeters of Regular Polygons (p.690)

____ **Goal:** Use a spreadsheet to explore the perimeters of regular polygons that are inscribed in a circle with a radius of one unit.

____ Student Help: Software Help

____ Other (specify) ______________________________

____

Homework and Additional Learning Support

____ Textbook (specify) pp. 686–689

____

____ Internet: Extra Examples at www.mcdougallittell.com

____ *Reteaching with Practice* worksheet (specify exercises)__________

____ *Personal Student Tutor* for Lesson 11.4

Lesson 11.4

NAME ______________________________ DATE __________

Real-Life Application: When Will I Ever Use This?

For use with pages 683–689

Running Track

The running track shown below has eight lanes. Each lane is 1.24 meters wide. There is a 180° arc at each end of the track and the straight sections have a length of 85 meters. The radii for the arcs in the first two lanes are given.

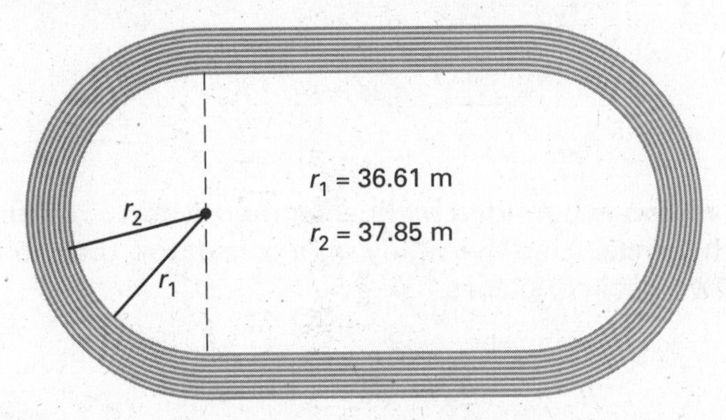

1. Find r_3, r_4, r_5, r_6, r_7, and r_8.
2. Find the distances around Lanes 1 through 8. Round your results to two decimal places.
3. Use your results from Exercise 2 to approximate the increase in distance between each lane.
4. Find the running distance between the straight aways for each lane. Round your results to two places.
5. Suppose you are racing two of your friends in the 200-meter dash. All three of you line up next to one another. What is wrong with your starting positions?

NAME ______________________________ DATE ________

Challenge: Skills and Applications

For use with pages 683–689

In Exercises 1–3, the region is bounded by line segments and circular arcs. Find the perimeter of the region. (Leave your answer in terms of π.)

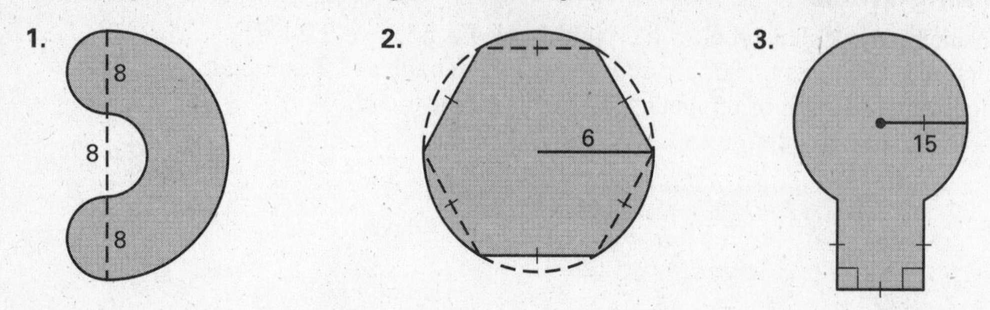

In Exercises 4–6, the region is bounded by line segments and circular arcs. Find the radius r of the circle. (Use the π key on a calculator, then round decimal answers to two decimal places.)

4. Perimeter = 12 **5.** Perimeter = 32 **6.** Perimeter = 25

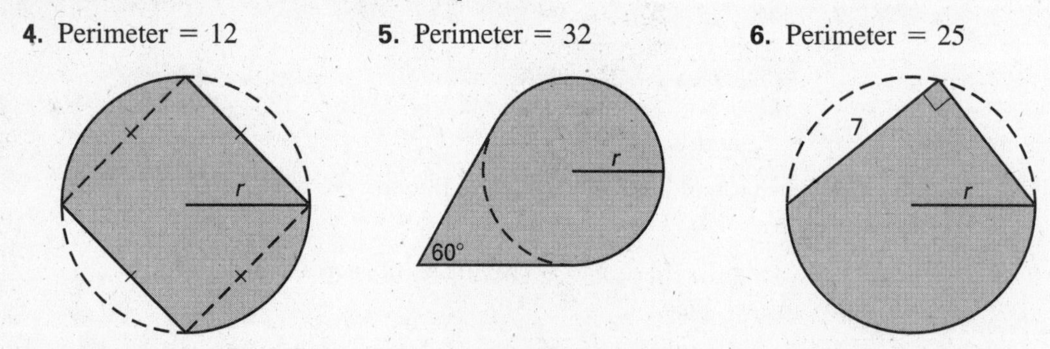

7. The Greek mathematician and astronomer Eratosthenes lived in the third century B.C. He is known for creating the "sieve of Eratosthenes," a method for finding prime numbers, and also for one of the most accurate early estimates of the radius of Earth.

Eratosthenes observed that at noon on the day of the summer solstice, the sun was shining directly down a deep well in Syene (S), indicating that the sun was directly overhead. At the same time in Alexandria (A), the sun was shining at an angle of 7.2°, as determined by measuring a shadow and using trigonometry. At that time in history, Alexandria was believed to be 500 miles due north of Syene.

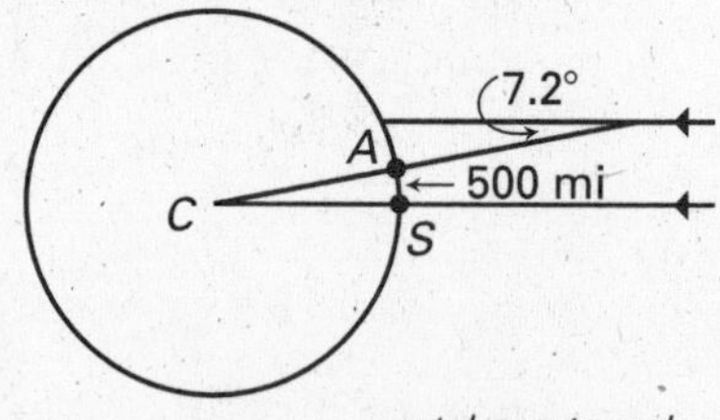

a. The sun's rays are parallel, and C represents the center of Earth. Find $m\angle ACS$. How do you know your answer is correct?

b. Find the circumference of Earth using Eratosthenes' measurements. Then find the radius.

c. The actual radius of Earth is about 3960 miles. How accurate was Eratosthenes' estimate?

LESSON 11.5

TEACHER'S NAME ______________________ CLASS ________ ROOM ________ DATE ________

Lesson Plan

1-day lesson (See *Pacing the Chapter,* TE pages 658C–658D)

For use with pages 691–698

GOALS
1. Find the area of a circle and a sector of a circle.
2. Use areas of circles and sectors to solve real-life problems.

State/Local Objectives ____

____

✓ Check the items you wish to use for this lesson.

STARTING OPTIONS

____ Homework Check: TE page 686: Answer Transparencies
____ Warm-Up or Daily Homework Quiz: TE pages 691 and 689, CRB page 69, or Transparencies

TEACHING OPTIONS

____ Motivating the Lesson: TE page 692
____ Lesson Opener (Application): CRB page 70 or Transparencies
____ Technology Activity with Keystrokes: CRB pages 71–74
____ Examples 1–6: SE pages 691–694
____ Extra Examples: TE pages 692–694 or Transparencies; Internet
____ Closure Question: TE page 694
____ Guided Practice Exercises: SE page 695

APPLY/HOMEWORK

Homework Assignment

____ Basic 10–28 even, 30–32, 35–37, 40–41, 43–44, 46–60 even
____ Average 10–28 even, 30–37, 40–41, 43–44, 46–60 even
____ Advanced 10–28 even, 30–37, 40–41, 43–45, 46–60 even

Reteaching the Lesson

____ Practice Masters: CRB pages 75–77 (Level A, Level B, Level C)
____ Reteaching with Practice: CRB pages 78–79 or Practice Workbook with Examples
____ Personal Student Tutor

Extending the Lesson

____ Applications (Interdisciplinary): CRB page 81
____ Challenge: SE page 698; CRB page 82 or Internet

ASSESSMENT OPTIONS

____ Checkpoint Exercises: TE pages 692–694 or Transparencies
____ Daily Homework Quiz (11.5): TE page 698, CRB page 85, or Transparencies
____ Standardized Test Practice: SE page 698; TE page 698; STP Workbook; Transparencies

Notes ____

____

____

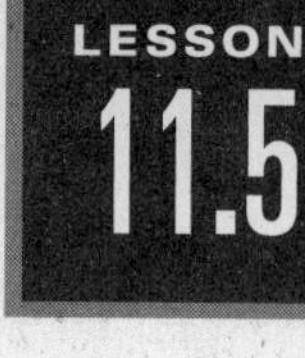

TEACHER'S NAME ____________________ CLASS ______ ROOM ______ DATE ______

Lesson Plan for Block Scheduling

Half-day lesson (See *Pacing the Chapter,* TE pages 658C–658D)

For use with pages 691–698

GOALS

1. **Find the area of a circle and a sector of a circle.**
2. **Use areas of circles and sectors to solve real-life problems.**

State/Local Objectives ____________________

____

____

CHAPTER PACING GUIDE	
Day	**Lesson**
1	11.1 (all)
2	11.2 (all)
3	11.3 (all)
4	11.4 (all)
5	**11.5 (all)**; 11.6 (begin)
6	11.6 (end); Review Ch. 11
7	Assess Ch. 11; 12.1 (all)

✓ Check the items you wish to use for this lesson.

STARTING OPTIONS

____ Homework Check: TE page 686: Answer Transparencies
____ Warm-Up or Daily Homework Quiz: TE pages 691 and 689, CRB page 69, or Transparencies

TEACHING OPTIONS

____ Motivating the Lesson: TE page 692
____ Lesson Opener (Application): CRB page 70 or Transparencies
____ Technology Activity with Keystrokes: CRB pages 71–74
____ Examples 1–6: SE pages 691–694
____ Extra Examples: TE pages 692–694 or Transparencies; Internet
____ Closure Question: TE page 694
____ Guided Practice Exercises: SE page 695

APPLY/HOMEWORK

Homework Assignment (See also the assignment for Lesson 11.6.)

____ Block Schedule: 10–28 even, 30–37, 40–41, 43–44, 46–60 even

Reteaching the Lesson

____ Practice Masters: CRB pages 75–77 (Level A, Level B, Level C)
____ Reteaching with Practice: CRB pages 78–79 or Practice Workbook with Examples
____ Personal Student Tutor

Extending the Lesson

____ Applications (Interdisciplinary): CRB page 81
____ Challenge: SE page 698; CRB page 82 or Internet

ASSESSMENT OPTIONS

____ Checkpoint Exercises: TE pages 692–694 or Transparencies
____ Daily Homework Quiz (11.5): TE page 698, CRB page 85, or Transparencies
____ Standardized Test Practice: SE page 698; TE page 698; STP Workbook; Transparencies

Notes ____________________

____

____

LESSON
11.5

NAME ______________________ DATE __________

Available as a transparency

WARM-UP EXERCISES

For use before Lesson 11.5, pages 691–698

Find the area of each figure.

1. a square with sides 12 in.

2. an equilateral triangle with sides 5 cm

3. a pentagon with apothem 2 ft and side length 2.9 ft

4. a hexagon with radius 16 yd

DAILY HOMEWORK QUIZ

For use after Lesson 11.4, pages 683–690

1. Find the circumference of a circle with a diameter 6 inches. (Use $\pi \approx 3.14$.)

2. Find the circumference of a circle with a radius 20 meters. (Leave your answer in terms of π.)

3. Find the length of $\widehat{AB}$. (Use $\pi \approx 3.14$.)

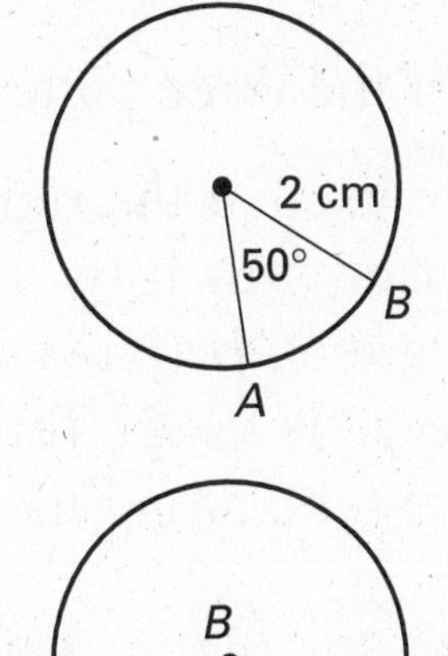

4. Find the circumference.

B
120°
C
A
25

5. Find the circumference of the circle whose equation is $x^2 + y^2 = 81$.

NAME ______________________ DATE __________

Available as a transparency

Application Lesson Opener

For use with pages 691–698

Sprinkler systems are designed so that all desired areas are watered completely and evenly. Sprinklers are placed so that the areas they water overlap, with at least two sprinklers covering every point.

Pop-up spray sprinklers are available in many patterns and sizes. The pattern is the shape of the area the sprinkler sprays. Each pattern shown below has a radius of 12 ft with water pressure of 20 psi.

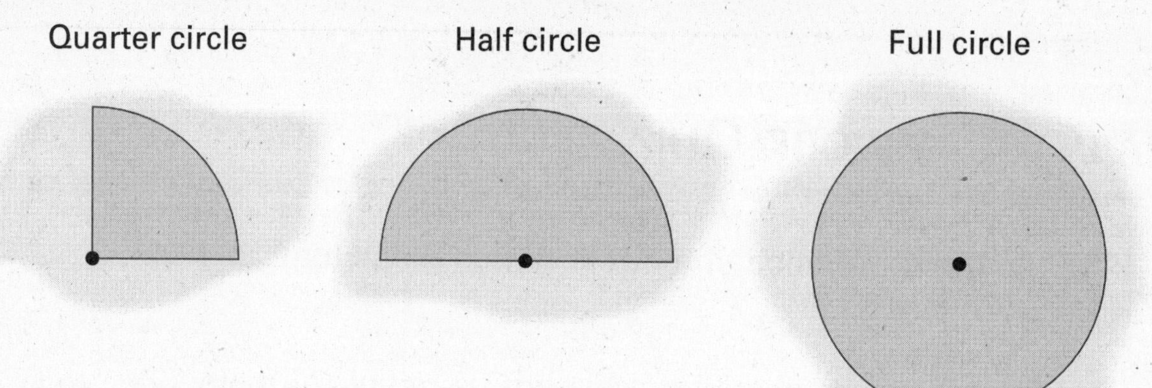

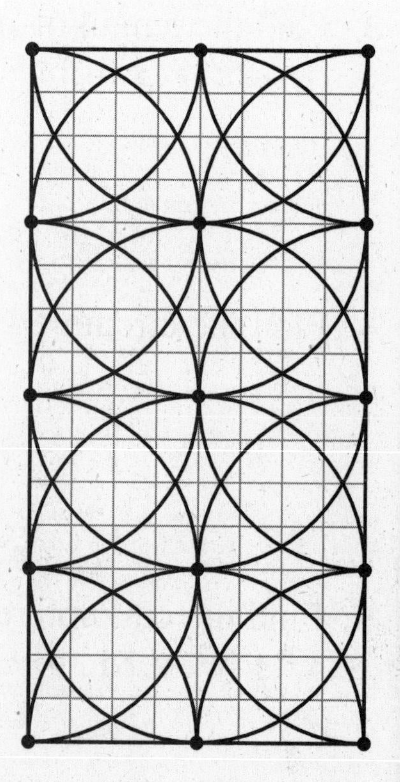

1. Compare the areas of the three patterns.
2. Fifteen sprinklers are used at the right. The rectangular area measures 24 ft by 48 ft. Label each sprinkler Q (quarter), H (half) or F (full). How many of each type is used? Tell which type is used at the corners, along the sides, and in the middle.
3. Draw a 12 ft-by-24 ft rectangular area covered by 6 sprinklers: 4 quarter circles and 2 half circles. Label each sprinkler.
4. Draw a 24-ft square area covered by 9 sprinklers: 4 quarter circles, 4 half circles, and 1 full circle. Label each sprinkler.

NAME ______________________ DATE ________

Technology Activity

For use with pages 691–698

GOAL **To determine the relationship that exists between a circle and special shaded regions in the circle**

Activity

1. Draw $\odot A$.
2. Draw a diameter in $\odot A$ and label the endpoints of the diameter B and C.
3. Locate the midpoints of radii $\overline{AB}$ and $\overline{AC}$; label the midpoints D and E, respectfully. Draw $\overline{AD}$ and $\overline{AE}$.
4. Draw $\odot D$ with radius AD, centered at D.
5. Draw $\odot E$ with radius AE, centered at E.
6. Measure the areas of $\odot A$, $\odot D$, and $\odot E$.
7. Measure the circumferences of $\odot A$, $\odot D$, and $\odot E$.
8. Use the calculate feature to divide the area of $\odot A$ by the sum of the areas of $\odot D$ and $\odot E$.
9. Use the calculate feature to divide the circumference of $\odot A$ by the sum of the circumferences of $\odot D$ and $\odot E$.
10. Change the size of $\odot A$ and observe the results.

Exercises

1. In Step 8 above, the area of $\odot A$ was shown to be twice that of the sum of $\odot D$ and $\odot E$. Show why this is true algebraically. (*Hint:* Let r be the radius of $\odot A$ and $\frac{1}{2}r$ be the radius for both $\odot D$ and $\odot E$.)
2. In Step 9, the circumference of $\odot A$ was shown to be the same as that of the sum of the circumferences of $\odot D$ and $\odot E$. Briefly explain why this is true.

NAME ______________________________ DATE __________

Technology Activity

For use with pages 691–698

TI-92

1. Draw ⊙A using the circle command (F3 1).
2. Draw a line that passes point A using the line command (F2 4). Draw a diameter in ⊙A and label the endpoints B and C.

   F2 5 (Move cursor to intersection of line and ⊙A.) ENTER B (Move cursor to other intersection of the line and ⊙A.) ENTER C
3. Locate the midpoints of radii $\overline{AB}$ and $\overline{AC}$ and label the midpoints D and E.

   F4 3 (Place cursor on A.) ENTER (Move cursor to B.) ENTER D

   Repeat this process for midpoint E of $\overline{AC}$. Then use the segment command (F2 5) to draw $\overline{AD}$ and $\overline{AE}$.
4. Draw ⊙D with radius $\overline{AD}$, centered at D.

   F4 8 (Place cursor on D.) ENTER (Place cursor on segment $\overline{AD}$.) ENTER
5. Draw ⊙E with radius $\overline{AE}$, centered at E. See Step 4.
6. Measure the areas of ⊙A, ⊙D, and ⊙E.

   F6 2 (Place cursor on ⊙A.) ENTER

   Repeat this process for ⊙D and ⊙E.
7. Measure the circumferences of ⊙A, ⊙D, and ⊙E.

   F6 1 (Place cursor on ⊙A.) ENTER

   Repeat this process for ⊙D and ⊙E.
8. Use the calculate feature to divide the area of ⊙A by the sum of the areas of ⊙D and ⊙E.

   F6 6 (Use cursor to highlight the area of ⊙A.) ENTER ÷ ((Highlight the area of ⊙D.) ENTER + (Highlight the area of ⊙E.) ENTER) ENTER (The result will appear on the screen.)

NAME ______________________ DATE ____________

Technology Activity

For use with pages 691–698

9. Use the calculate feature to divide the circumference of $\odot A$ by the sum of the circumferences of $\odot D$ and $\odot E$.

F6 6 (Use cursor to highlight the circumference of $\odot A$.) ENTER ÷ ((Highlight the circumference of $\odot D$.) ENTER + (Highlight the circumference of $\odot E$.) ENTER) ENTER (The result will appear on the screen.)

10. Drag $\odot A$ and observe the results.

F1 1 (Move cursor to edge of $\odot A$.)

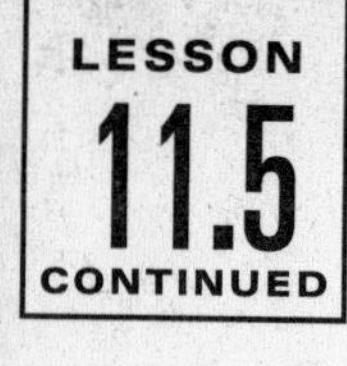

NAME _______________________________ DATE ____________

Technology Activity

For use with page 691–698

SKETCHPAD

1. Draw ⊙A using the compass tool.
2. Draw a diameter in ⊙A and label the endpoints of the diameter B and C. Select the line straightedge tool, place the cursor on A and move the line so that it passes through B. Choose the point tool and label the intersection of ⊙A and the line as C. Choose the segment straightedge tool and draw $\overline{BC}$, $\overline{AB}$, and $\overline{AC}$.
3. Locate the midpoints of radii $\overline{AB}$ and $\overline{AC}$ and label the midpoints D and E, respectively. Use the selection arrow tool to select $\overline{AB}$. Choose **Point at Midpoint** from the **Construct** menu. Repeat for $\overline{AC}$. Use the text tool to label the points, if necessary. Choose the segment straightedge tool and draw $\overline{AE}$ and $\overline{AD}$.
4. Draw ⊙D with radius AD, centered at D. Use the selection arrow tool to select point D and $\overline{AD}$. From the **Construct** menu, choose **Circle by Center and Radius**.
5. Draw ⊙E with radius AE, centered at E. See Step 4.
6. Measure the areas of ⊙A, ⊙D, and ⊙E. Use the selection arrow tool to select ⊙A. From the **Measure** menu, choose **Area**. Repeat this process for ⊙D and ⊙E.
7. Measure the circumferences of ⊙A, ⊙D, and ⊙E. Use the selection arrow tool to select ⊙A. From the **Measure** menu, choose **Circumference**. Repeat this process for ⊙D and ⊙E.
8. Use the calculate feature to divide the area of ⊙A by the sum of the areas of ⊙D and ⊙E. From the **Measure** menu, choose **Calculate**. Click the area of ⊙A, click [/], click [( ], click the area of ⊙D, click [+], click the area of ⊙E, click [ )], and click OK.
9. Use the calculate feature to divide the circumference of ⊙A by the sum of the circumferences of ⊙D and ⊙E. From the **Measure** menu, choose **Calculate**. Click the circumference of ⊙A, click [/], click [( ], click the circumference of ⊙D, click [+], click the circumference of ⊙E, click [ )], and click OK.
10. Use the translate selection arrow tool to select point A and change the sizes of the circles.

LESSON

11.5

NAME ______________________ DATE __________

Practice A

For use with pages 691–698

Match the measure with its value.

1. $m\widehat{AB}$
2. Area $\odot C$
3. Area of shaded region I
4. Area of shaded region II
5. Area of unshaded region

A. 2π units^2

B. 16π units^2

C. 6π units^2

D. $180°$

E. 8π units^2

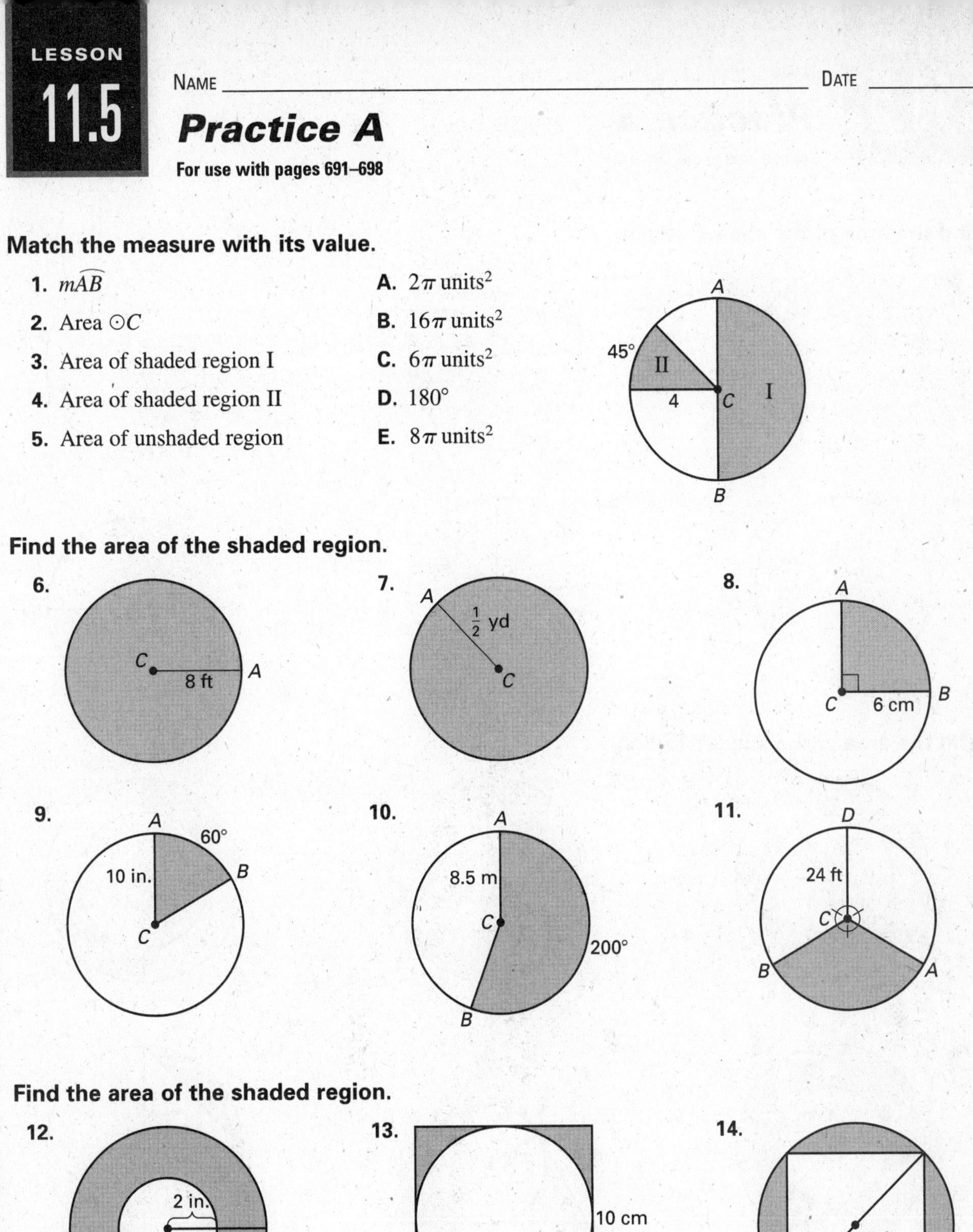

Find the area of the shaded region.

6. C, 8 ft, A

7. A, $\frac{1}{2}$ yd, C

8. A, C, 6 cm, B

9. A, 60°, 10 in., B, C

10. A, 8.5 m, C, 200°, B

11. D, 24 ft, C, B, A

Find the area of the shaded region.

12. 2 in., 4 in.

13.

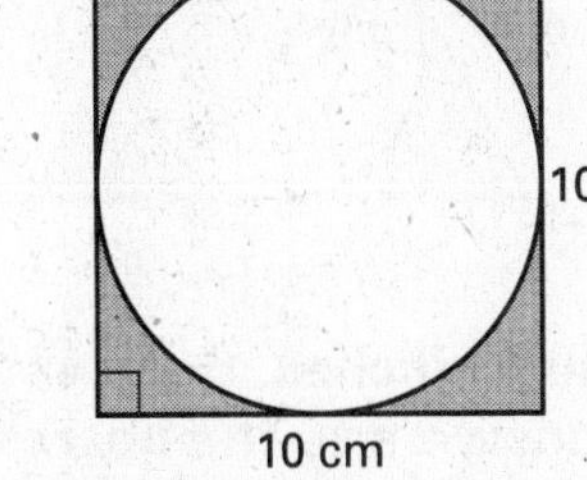

14.

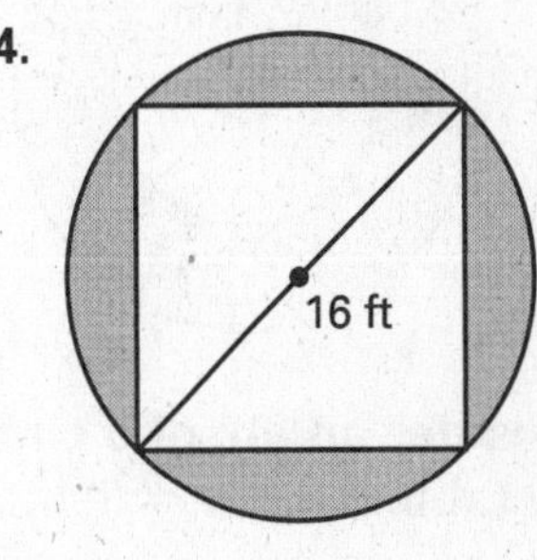

Determine the radius of the circle with the given area.

15. $A = 25\pi$ cm^2

16. $A = 144\pi$ in.^{2}

17. $A = 48$ ft^2

NAME ______________________________ DATE ____________

Practice B

For use with pages 691–698

Find the area of the shaded region.

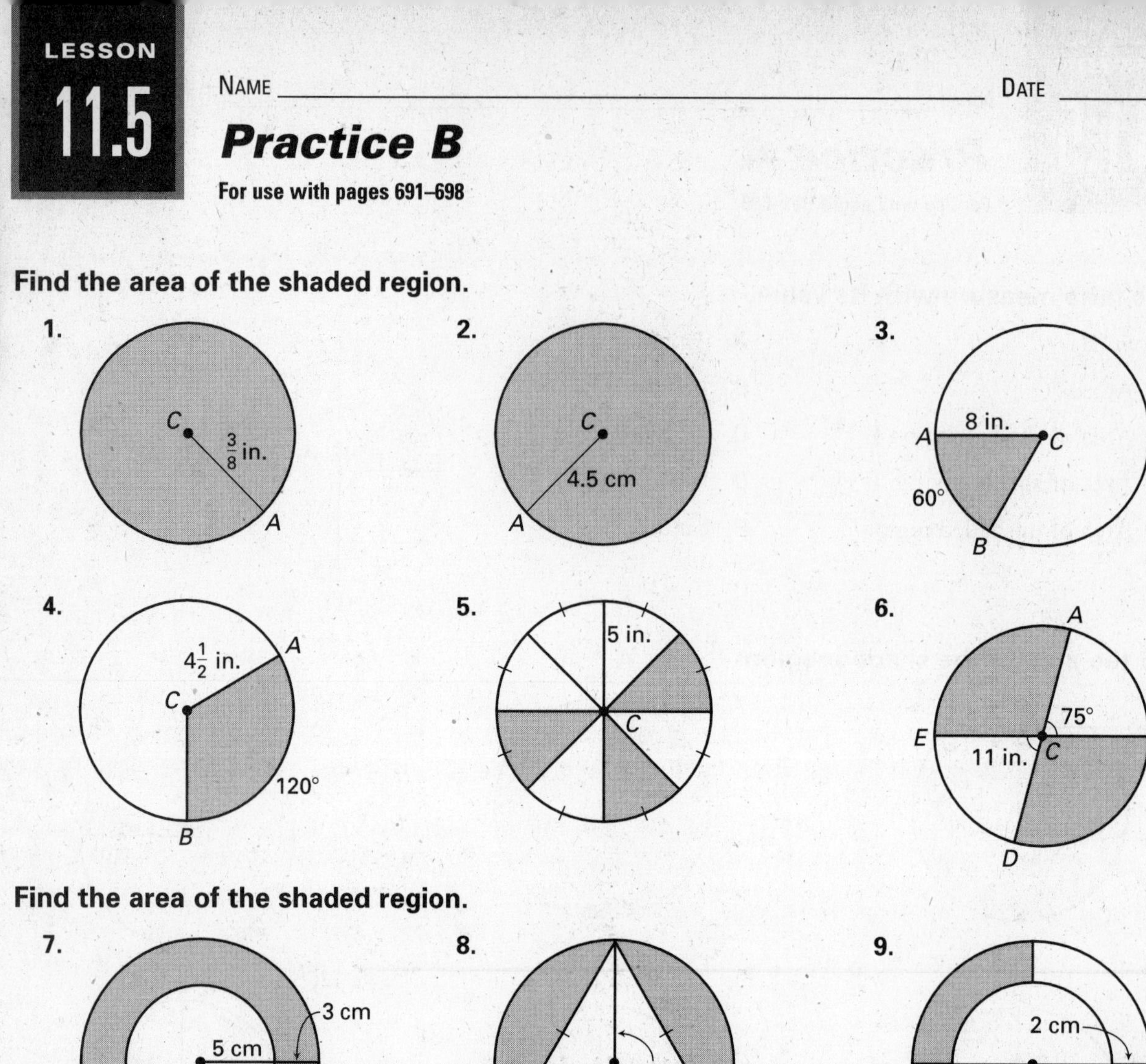

Find the area of the shaded region.

7.

3 cm
5 cm

8.

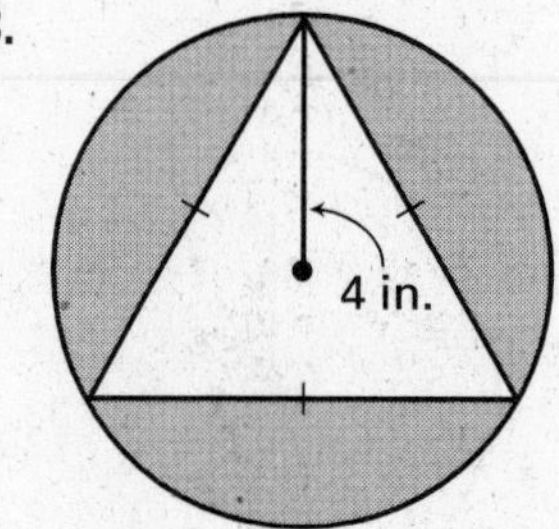

9.

2 cm
8 cm

10.

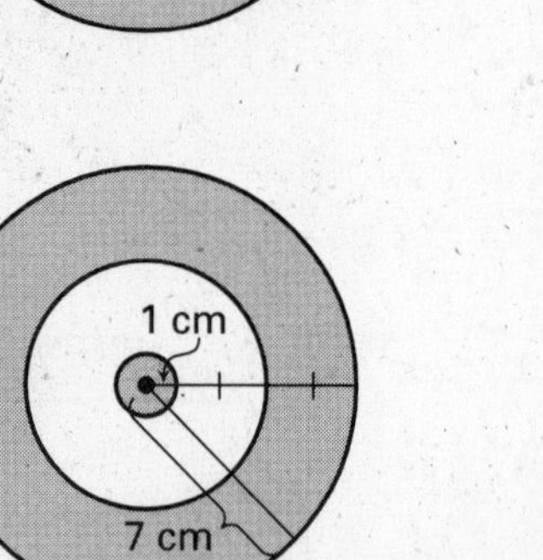

11.

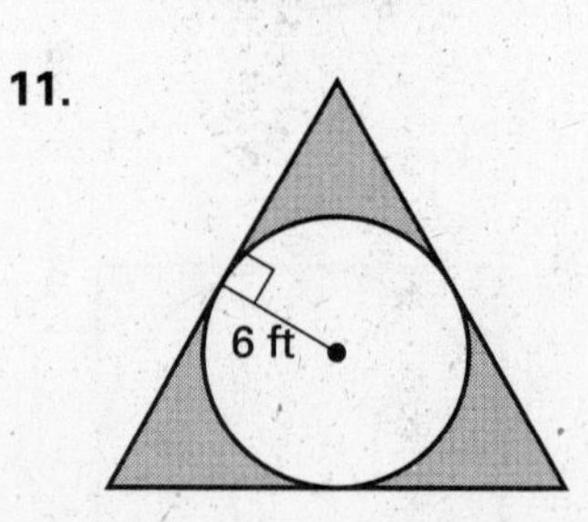

12.

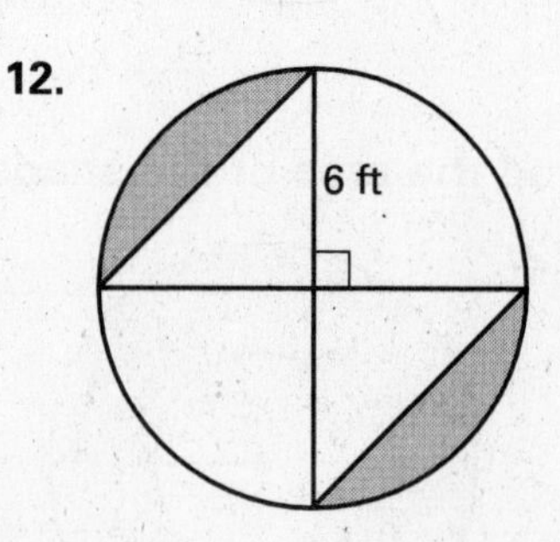

Consider an arc of a circle with radius 4 inches. Copy and complete the table. Use $\pi \approx 3.14$ and round answers to the nearest tenth.

13.

Measure of arc	30°	60°	90°	120°	150°	180°
Area of corresponding sector						

LESSON 11.5

NAME ______________________ DATE __________

Practice C

For use with pages 691–698

Find the area of the shaded region.

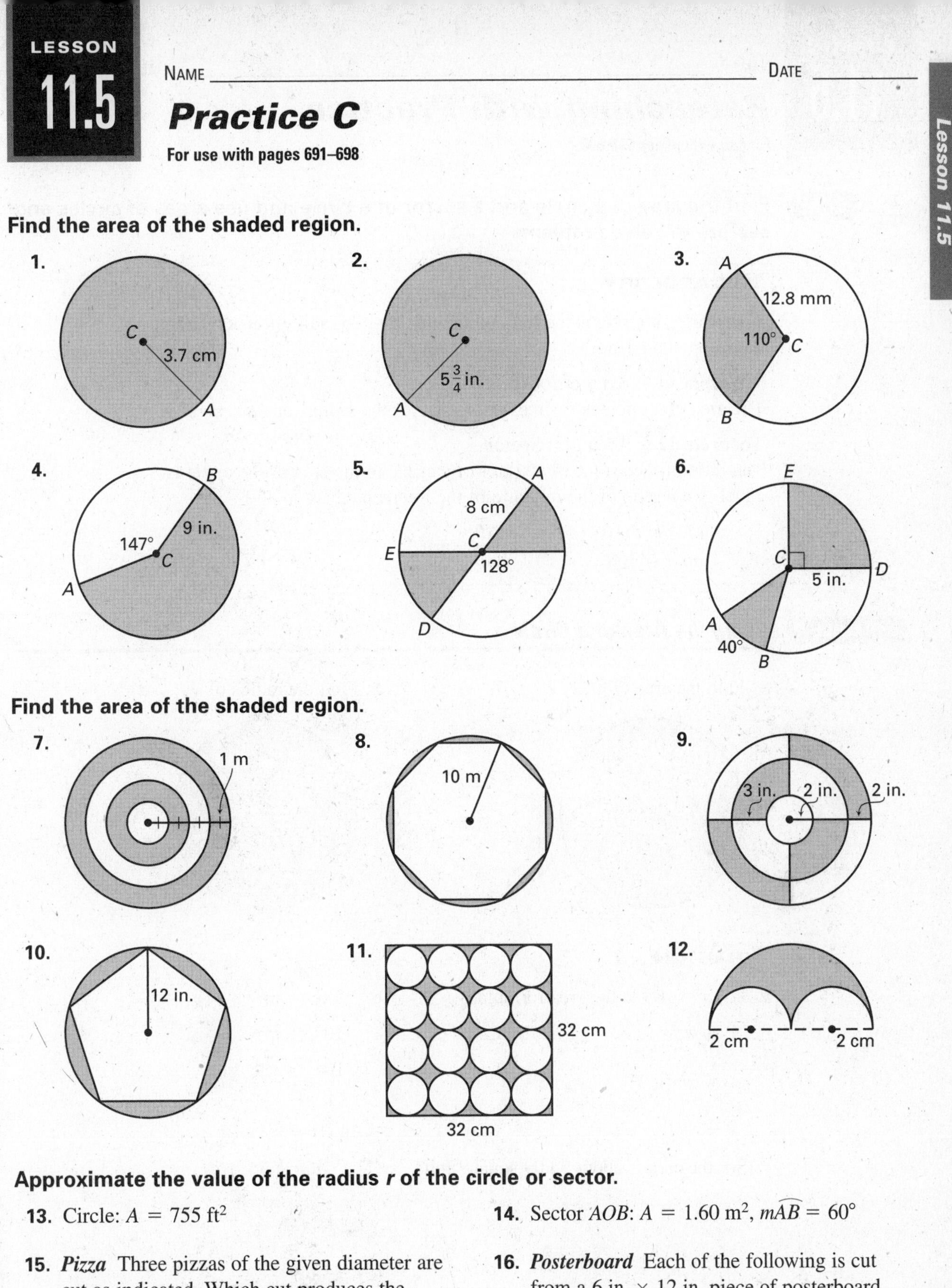

Find the area of the shaded region.

Approximate the value of the radius *r* of the circle or sector.

13. Circle: $A = 755\text{ ft}^2$

14. Sector AOB: $A = 1.60\text{ m}^2$, $m\widehat{AB} = 60°$

15. ***Pizza*** Three pizzas of the given diameter are cut as indicated. Which cut produces the largest pieces?

a. An 8-inch pizza cut into 6 congruent slices

b. A 12-inch pizza cut into 8 congruent slices

c. A 16-inch pizza cut into 10 congruent slices

16. ***Posterboard*** Each of the following is cut from a 6 in. × 12 in. piece of posterboard. Which wastes the least?

a. Two 6-inch diameter circles

b. Eight 3-inch diameter circles

c. Eighteen 2-inch diameter circles

NAME ______________________ DATE ________

Reteaching with Practice

For use with pages 691–698

GOAL **Find the area of a circle and a sector of a circle and use areas of circles and sectors to solve problems**

VOCABULARY

A **sector of a circle** is the region bounded by two radii of the circle and their intercepted arc.

Theorem 11.7 Area of a Circle
The area of a circle is π times the square of the radius, or $A = \pi r^2$.

Theorem 11.8 Area of a Sector
The ratio of the area A of a sector of a circle to the area of the circle is equal to the ratio of the measure of the intercepted arc to 360°.

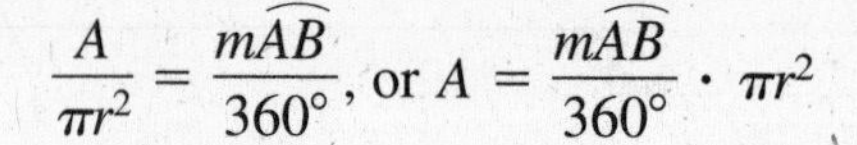

$$\frac{A}{\pi r^2} = \frac{m\widehat{AB}}{360^\circ}, \text{ or } A = \frac{m\widehat{AB}}{360^\circ} \cdot \pi r^2$$

EXAMPLE 1 ***Using the Area of a Circle***

a. Find the area of $\odot C$.

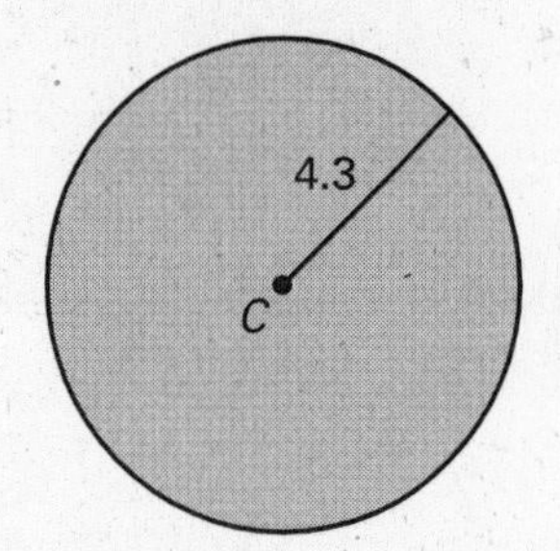

b. Find the radius of $\odot P$.

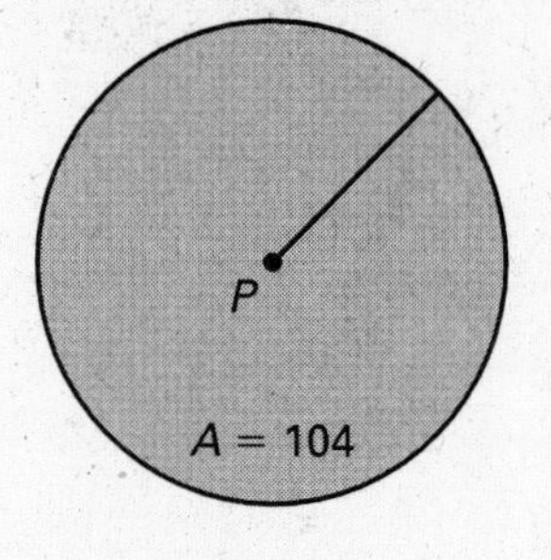

SOLUTION

a. Use $r = 4.3$ in the area formula.

$A = \pi r^2$

$A = \pi \cdot 4.3^2$

$A \approx 58.09$

So, the area is about 58.09 square units.

b.

$A = \pi r^2$

$104 = \pi r^2$

$\frac{104}{\pi} = r^2$

$33.10 \approx r^2$

$r \approx 5.75$

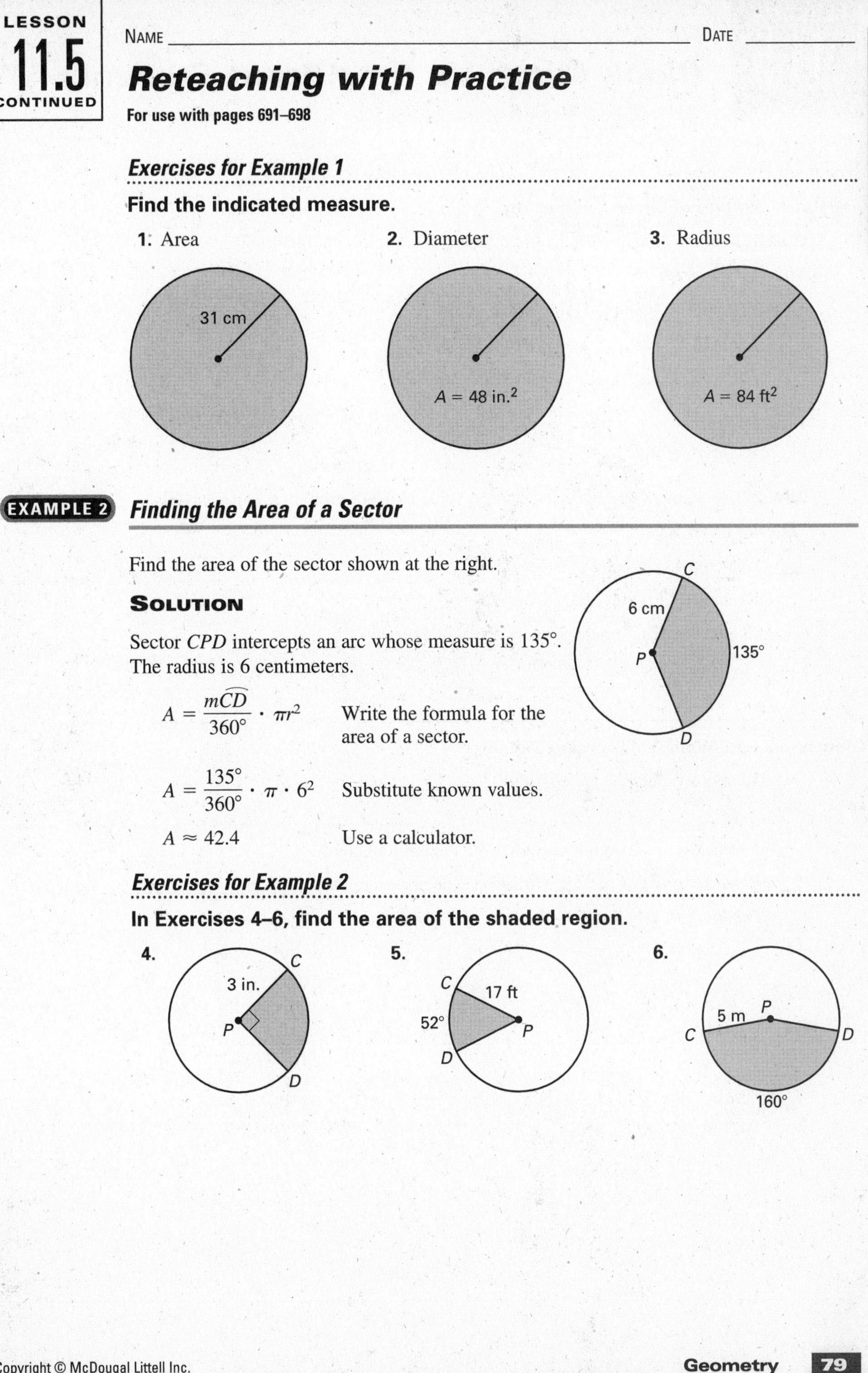

Reteaching with Practice

For use with pages 691–698

Exercises for Example 1

Find the indicated measure.

1. Area

2. Diameter

3. Radius

EXAMPLE 2 *Finding the Area of a Sector*

Find the area of the sector shown at the right.

SOLUTION

Sector *CPD* intercepts an arc whose measure is 135°. The radius is 6 centimeters.

$A = \frac{m\overset{\frown}{CD}}{360^\circ} \cdot \pi r^2$ Write the formula for the area of a sector.

$A = \frac{135^\circ}{360^\circ} \cdot \pi \cdot 6^2$ Substitute known values.

$A \approx 42.4$ Use a calculator.

Exercises for Example 2

In Exercises 4–6, find the area of the shaded region.

4.

5.

6.

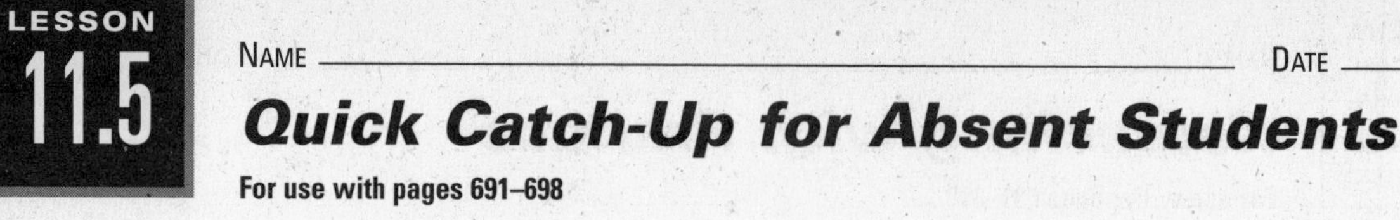

Quick Catch-Up for Absent Students

For use with pages 691–698

The items checked below were covered in class on (date missed) ____________

Lesson 11.5: Areas of Circles and Sectors

____ **Goal 1:** Find the area of a circle and a sector of a circle. (pp. 691–692)

Material Covered:

____ Example 1: Using the Area of a Circle

____ Example 2: Finding the Area of a Sector

____ Example 3: Finding the Area of a Sector

Vocabulary:

sector of a circle, p. 692

____ **Goal 2:** Use areas of circles and sectors to solve real-life problems. (pp. 693–694)

Material Covered:

____ Example 4: Finding the Area of a Region

____ Example 5: Finding the Area of a Region

____ Example 6: Finding the Area of a Boomerang

____ Other (specify) ____

____

Homework and Additional Learning Support

____ Textbook (specify) pp. 695–698

____

____ Internet: Extra Examples at www.mcdougallittell.com

____ *Reteaching with Practice* worksheet (specify exercises)____________

____ *Personal Student Tutor* for Lesson 11.5

NAME ______________________________ DATE __________

Interdisciplinary Application

For use with pages 691–698

Meteorology

EARTH SCIENCE Meteorology is the study of Earth's atmospheric conditions and how those conditions affect the weather we experience. Although there are several types of meteorologists, including observers, forecasters, and researchers, they all have the common goal of trying to accurately predict the weather. This objective is extremely important when it comes to warning people about dangerous storms, such as tornadoes and hurricanes.

One very important tool of weather forecasting is the radar. It is often capable of detecting not only precipitation but also the strength, the course, and the speed of a storm within a 250 mile radius. It is not only common areas of extreme weather that rely on radar, but also most major airports. Passengers on flights are much safer when flights can be rerouted to avoid dangerous weather conditions.

1. Your class goes on a field trip to a weather observatory. A meteorologist is setting up a radar station. The meteorologist says they are tracking a particular storm approaching from the southeast. The radar is set to reach a distance of 200 miles and cover an arc of 150°. Find the area of the sector the radar covers.

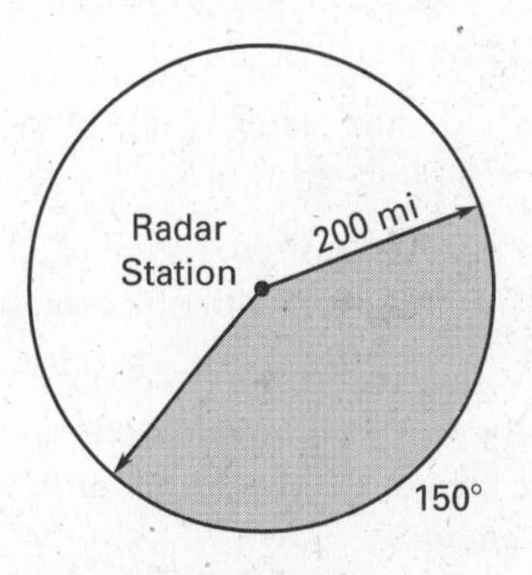

2. A second radar station is set to reach a distance of 75 miles and cover a complete circle. Draw a diagram to model the situation. Find the area of the circle the radar covers.

3. A radar station is set up to observe the weather activity on the waters of Lake Erie. This station will reach a distance of 60 miles and cover a 180° arc. Draw a diagram to model the situation. Find the area of the sector the radar will cover.

NAME _______________ DATE _______

Challenge: Skills and Applications

For use with pages 691–698

In Exercises 1–3, find the area of the shaded region. (Leave your answer in terms of π. Note that in Exercise 3, the center of each circle is a vertex of the regular hexagon.)

1. **2.** **3.**

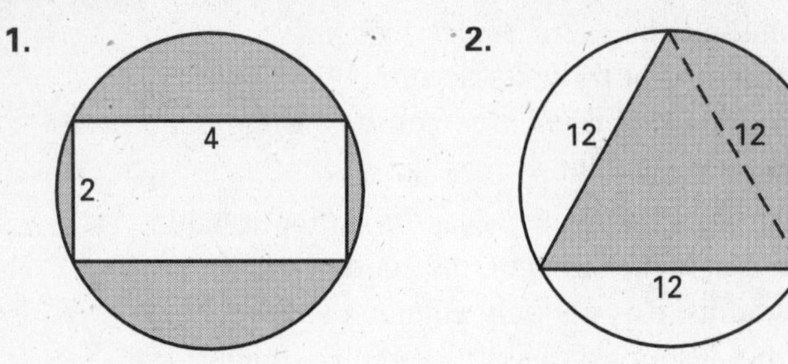

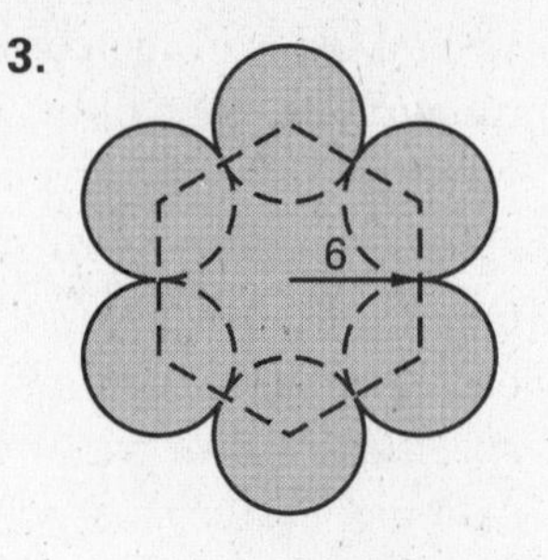

4. The diagram shows a region whose boundary consists of three congruent arcs of measure 270° and three congruent line segments.

a. If r is the radius of one of the arcs, find the area of the region in terms of r.

b. If the area of the region is 30 cm^2, what is r? Round to the nearest hundredth of a centimeter.

5. The diagram shows a rectangle inscribed in a circle, as well as 4 semicircles whose diameters are sides of the rectangle. Prove that the area of the shaded region is ab. (*Hint:* Use the Pythagorean theorem.)

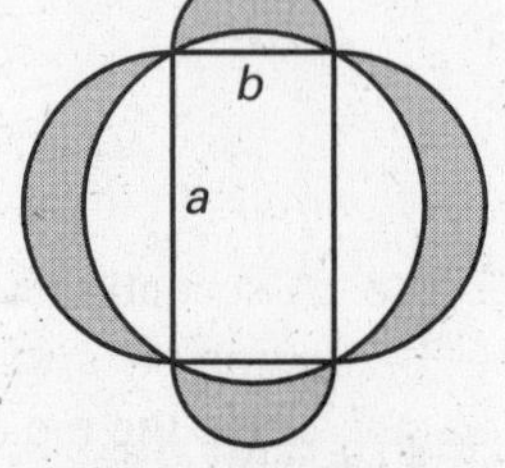

6. In the diagram, all four shaded regions have the same area. If the smallest circle has radius 1, find the radii of the other three circles.

7. In the diagram, $\widehat{ABE}$ is a semicircle with center C, and $\widehat{ADB}$ is a semicircle with center F. If $AE = 12$ in., find the area of the shaded region.

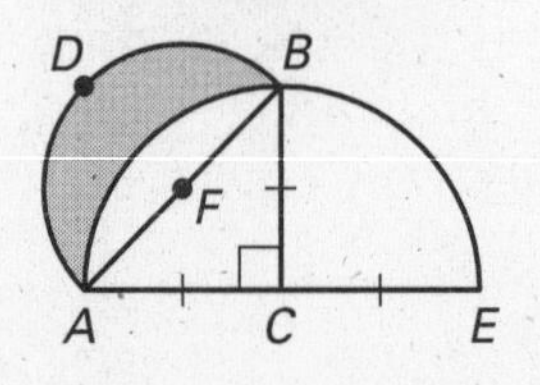

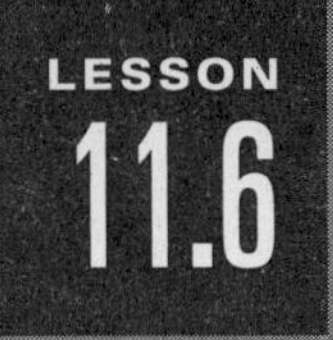

LESSON 11.6

TEACHER'S NAME ______________________ CLASS ______ ROOM ______ DATE ______

Lesson Plan

2-day lesson (See *Pacing the Chapter,* TE pages 658C–658D)

For use with pages 699–706

GOALS

1. **Find a geometric probability.**
2. **Use geometric probability to solve real-life problems.**

State/Local Objectives ____

____

✓ Check the items you wish to use for this lesson.

STARTING OPTIONS

____ Homework Check: TE page 695: Answer Transparencies
____ Warm-Up or Daily Homework Quiz: TE pages 699 and 698, CRB page 85, or Transparencies

TEACHING OPTIONS

____ Lesson Opener (Activity): CRB page 86 or Transparencies
____ Technology Activity with Keystrokes: CRB pages 87–88
____ Examples: Day 1: 1–2, SE pages 699–700; Day 2: 3–4, SE pages 700–701
____ Extra Examples: Day 1: TE page 700 or Transp.; Day 2: TE pages 700–701 or Transp.; Internet
____ Technology Activity: SE page 706
____ Closure Question: TE page 701
____ Guided Practice: SE page 701 Day 1: Exs. 1–3; Day 2: Exs. 4–8

APPLY/HOMEWORK

Homework Assignment

____ Basic Day 1: 9–25; Day 2: 26–34, 36–39, 43, 45–52; Quiz 2: 1–7
____ Average Day 1: 9–25; Day 2: 26–43, 45–52; Quiz 2: 1–7
____ Advanced Day 1: 9–25; Day 2: 26–52; Quiz 2: 1–7

Reteaching the Lesson

____ Practice Masters: CRB pages 89–91 (Level A, Level B, Level C)
____ Reteaching with Practice: CRB pages 92–93 or Practice Workbook with Examples
____ Personal Student Tutor

Extending the Lesson

____ Applications (Real-Life): CRB page 95
____ Challenge: SE page 704; CRB page 96 or Internet

ASSESSMENT OPTIONS

____ Checkpoint Exercises: Day 1: TE page 700 or Transp.; Day 2: TE pages 700–701 or Transp.
____ Daily Homework Quiz (11.6): TE page 704, or Transparencies
____ Standardized Test Practice: SE page 704; TE page 704; STP Workbook; Transparencies
____ Quiz (11.4–11.6): SE page 705; CRB page 97

Notes ____

____

____

LESSON 11.6

TEACHER'S NAME ______________________ CLASS ________ ROOM ________ DATE ________

Lesson Plan for Block Scheduling

1-day lesson (See *Pacing the Chapter,* TE pages 658C–658D) **For use with pages 699–706**

GOALS

1. Find a geometric probability.
2. Use geometric probability to solve real-life problems.

State/Local Objectives ______________________________

____

____

CHAPTER PACING GUIDE	
Day	**Lesson**
1	11.1 (all)
2	11.2 (all)
3	11.3 (all)
4	11.4 (all)
5	11.5 (all); **11.6 (begin)**
6	**11.6 (end)**; Review Ch. 11
7	Assess Ch. 11; 12.1 (all)

✓ Check the items you wish to use for this lesson.

STARTING OPTIONS

____ Homework Check: TE page 695: Answer Transparencies
____ Warm-Up or Daily Homework Quiz: TE pages 699 and 698, CRB page 85, or Transparencies

TEACHING OPTIONS

____ Lesson Opener (Activity): CRB page 86 or Transparencies
____ Technology Activity with Keystrokes: CRB pages 87–88
____ Examples: Day 5: 1–2, SE pages 699–700; Day 6: 3–4, SE pages 700–701
____ Extra Examples: Day 5: TE page 700 or Transp.; Day 6: TE pages 700–701 or Transp.; Internet
____ Technology Activity: SE page 706
____ Closure Question: TE page 701
____ Guided Practice: SE page 701 Day 5: Exs. 1–3; Day 6: Exs. 4–8

APPLY/HOMEWORK

Homework Assignment (See also the assignment for Lesson 11.5.)

____ Block Schedule: Day 5: 9–25; Day 6: 26–43, 45–52; Quiz 2: 1–7

Reteaching the Lesson

____ Practice Masters: CRB pages 89–91 (Level A, Level B, Level C)
____ Reteaching with Practice: CRB pages 92–93 or Practice Workbook with Examples
____ Personal Student Tutor

Extending the Lesson

____ Applications (Real-Life): CRB page 95
____ Challenge: SE page 704; CRB page 96 or Internet

ASSESSMENT OPTIONS

____ Checkpoint Exercises: Day 5: TE page 700 or Transp.; Day 6: TE pages 700–701 or Transp.
____ Daily Homework Quiz (11.6): TE page 704, or Transparencies
____ Standardized Test Practice: SE page 704; TE page 704; STP Workbook; Transparencies
____ Quiz (11.4–11.6): SE page 705; CRB page 97

Notes ______________________________

____

____

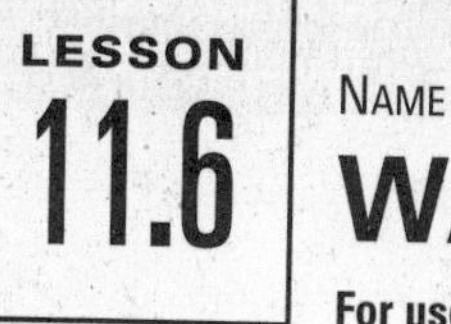

NAME ______________________ DATE __________

Available as a transparency

WARM-UP EXERCISES

For use before Lesson 11.6, pages 699–706

Find the area of each figure.

1. $\odot R$
2. hexagon *ABCDEF*
3. the shaded region

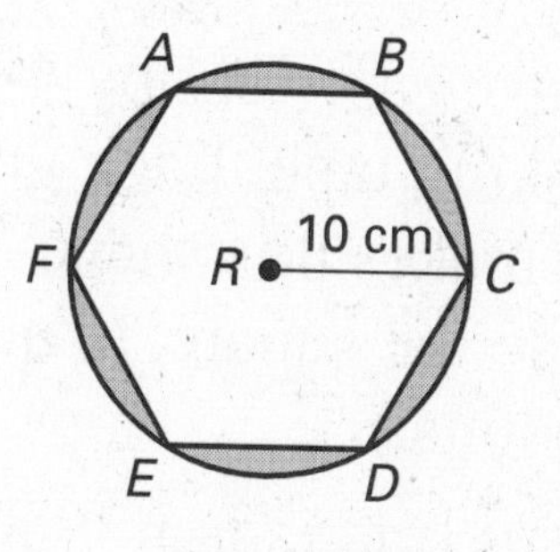

DAILY HOMEWORK QUIZ

For use after Lesson 11.5, pages 691–698

1. What is the area of a circle with radius 9 cm?
2. Find the area of the shaded region.

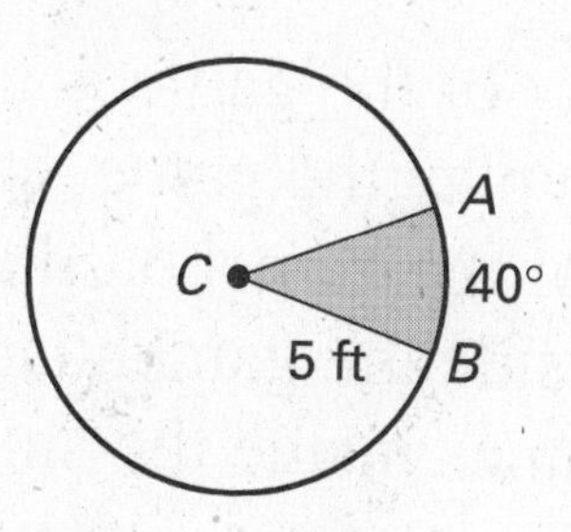

3. Find the area of the shaded region.

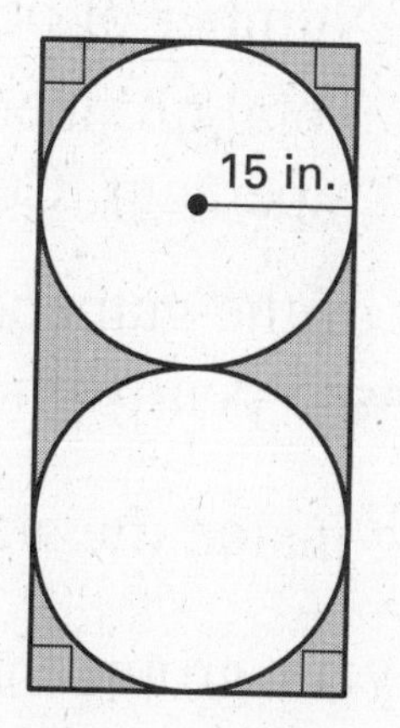

4. What is the radius of a circle with area 100π square meters?

Lesson 11.6

NAME ______________________ DATE __________

Available as a transparency

Activity Lesson Opener

For use with pages 699–705

SET UP: Work in a group.
YOU WILL NEED: • measuring tape • masking tape • pennies

1. Each group makes a 4-foot by 4-foot square gameboard on the floor with masking tape. Each group chooses a shape from the list below and makes this shape on the floor with masking tape anywhere inside the square. (Your shape can touch the edge at one or more points.)

   a. a square with side length 2 feet

   b. a rectangle with length 16 inches and width 36 inches

   c. a rectangle with length 4 feet and width 1 foot

   d. an isosceles right triangle with hypotenuse 4 feet

   e. an equilateral triangle with side length 3 feet

2. The object of the game is to toss pennies that land inside the shape. To play the game, each member of the group stands at an edge of the gameboard and tosses 10 pennies. If a penny lands outside the gameboard, try again. The person with the most pennies inside the shape wins. Do not move the pennies.

3. Find the ratio $\frac{\text{Number of pennies inside shape}}{\text{Total number of pennies tossed}}$ for your group and write it as a percent. Compare results with other groups.

4. If every point of the gameboard were an equally likely landing point for a tossed penny, then the *geometric probability* of a penny landing inside the shape is $\frac{\text{Area of shape}}{\text{Area of game board}}$. Find this ratio for your group and write it as a percent. Compare with your result in Exercise 3 and make a conjecture or a comment. Compare with other groups and examine their gameboards.

NAME ______________________ DATE __________

Technology Activity Keystrokes

For use with page 706

SHARP EL-9600c

```
DARTS
Print "HOW MANY THROWS
Input N
0⇒H
0⇒I
Label 1
random⇒X
random⇒Y
If (X²+Y²)<0.25 Goto 2
I+1⇒I
Goto 3
Label 2
H+1⇒H
I+1⇒I
Goto 3
Label 3
If I=N Goto 4
Goto 1
Label 4
Print "NUMBER OF HITS
Print H
End
```

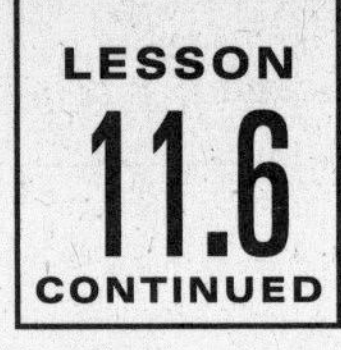

NAME ______________________ DATE ________

Technology Activity Keystrokes

For use with page 706

Lesson 11.6

CASIO CFX-9850GA PLUS

DARTS

"HOW MANY THROWS"? → H↵

0 → H↵

For 1 → I TO N↵

Ran# → X↵

Ran# → Y↵

If $(X^2+Y^2)<0.25$↵

Then H+1 → H↵

IfEnd↵

Next↵

"NUMBER OF HITS"↵

H↵

NAME ______________________________ DATE __________

Practice C

For use with pages 699–705

Find the probability that a point K, selected randomly on $\overline{AF}$, is on the given segment.

1. $\overline{AB}$
2. $\overline{CD}$
3. $\overline{BD}$
4. $\overline{CF}$

A B C D E F
−15 −9 −3 3 9 15 21

A point is chosen on $\overline{PQ}$. Determine the probability described.

5. The point is closer to point L than to point P.
6. The point is closer to point L than to point Q.

0 2 4 6 8 10 12
P L Q

Find the probability that a randomly chosen point in the figure lies in the shaded region.

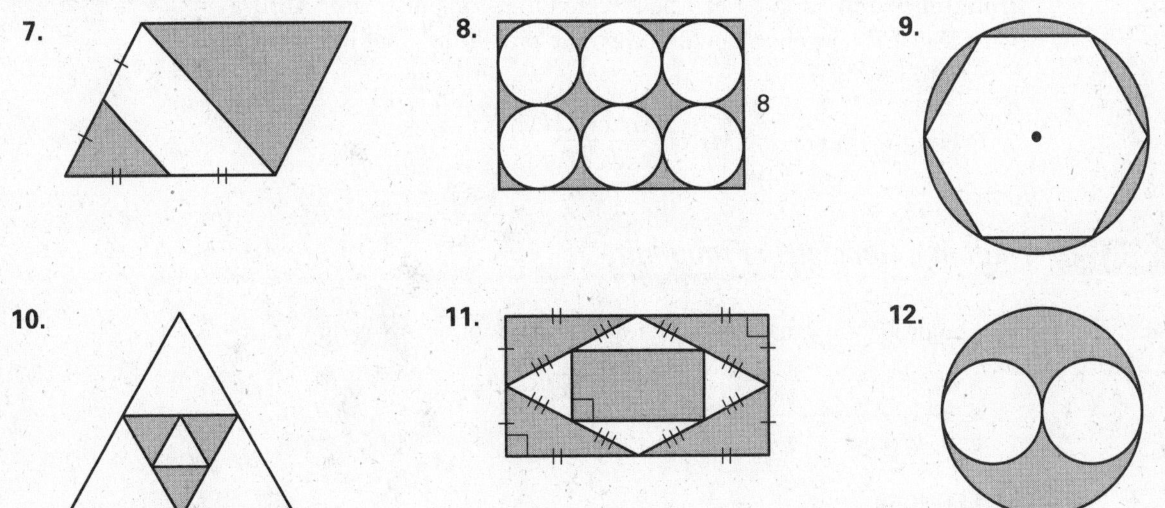

13. *City Bus* You have planned to meet your friend at the mall at 4 P.M. The city bus runs every 10 minutes and the trip to the store is 6 minutes. You arrive at the bus stop at 3:50 P.M. What is the probability that you will arrive at the mall by 4 P.M.?

14. *Coffee* You stop at the same convenience store each day to get a refill of your travel mug. The coffee decanter holds 64 ounces and your mug holds 20 ounces. What is the probability that on any given day you will have to tell the store manager that the coffee is out and they need to make more for you to fill your entire mug?

15. *Archery Target* Determine the probability for each outcome on the archery target shown. The center ring has a radius of 2 units. Each successive ring has a radius 1 unit greater than the previous one. Assume the arrow is equally likely to hit any point on the target.

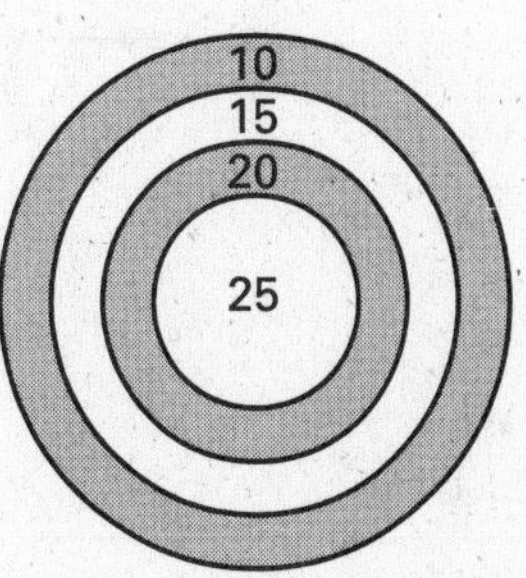

a. 25 points b. 20 points

c. 15 points d. 10 points

NAME ____________________ DATE __________

Reteaching with Practice

For use with pages 699–705

GOAL Find a geometric probability

VOCABULARY

A **probability** is a number from 0 to 1 that represents the chance that an event will occur.

Geometric probability is a probability that involves a geometric measure such as length or area.

Probability and Length Let $\overline{AB}$ be a segment that contains the segment $\overline{CD}$. If a point K on $\overline{AB}$ is chosen at random, then the probability that it is on $\overline{CD}$ is as follows:

$$P(\text{Point } K \text{ is on } \overline{CD}) = \frac{CD}{AB} = \frac{\text{Length of } \overline{CD}}{\text{Length of } \overline{AB}}$$

Probability and Area Let J be a region that contains region M. If a point K in J is chosen at random, then the probability that it is in region M is as follows:

$$P(\text{Point } K \text{ is in region } M) = \frac{\text{Area of } M}{\text{Area of } J}$$

EXAMPLE 1 Finding a Geometric Probability

Find the probability that a point chosen at random on $\overline{AB}$ is on $\overline{CD}$.

A C D B

0 1 2 3 4 5 6 7 8 9 10 11 12

SOLUTION

$$P(\text{Point is on } \overline{CD}) = \frac{\text{Length of } \overline{CD}}{\text{Length of } \overline{AB}} = \frac{8}{12} = \frac{2}{3}$$

The probability can be written as $\frac{2}{3}$, or approximately 0.667, or 66.7%.

Exercises for Example 1

In Exercises 1–4, find the probability that a point A, selected randomly on $\overline{AB}$, is on the given segment.

A C E D F B

0 1 2 3 4 5 6 7 8

1. $\overline{CD}$
2. $\overline{EF}$
3. $\overline{CF}$
4. $\overline{CE}$

Reteaching with Practice

For use with pages 699–705

EXAMPLE 2 Using Areas to Find a Geometric Probability

Find the probability that a point chosen at random in parallelogram *ABCD* lies in the shaded region.

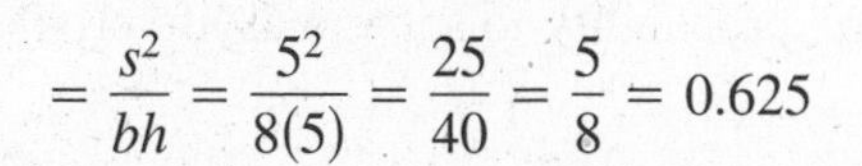

SOLUTION

Find the ratio of the area of the shaded square to the area of the parallelogram.

$$P(\text{point is in shaded region}) = \frac{\text{Area of shaded region}}{\text{Area of parallelogram}}$$

$$= \frac{s^2}{bh} = \frac{5^2}{8(5)} = \frac{25}{40} = \frac{5}{8} = 0.625$$

The probability that a point chosen at random in parallelogram *ABCD* lies in the square is 0.625.

Exercises for Example 2

Find the probability that a point chosen at random in the figure lies in the shaded region.

5.

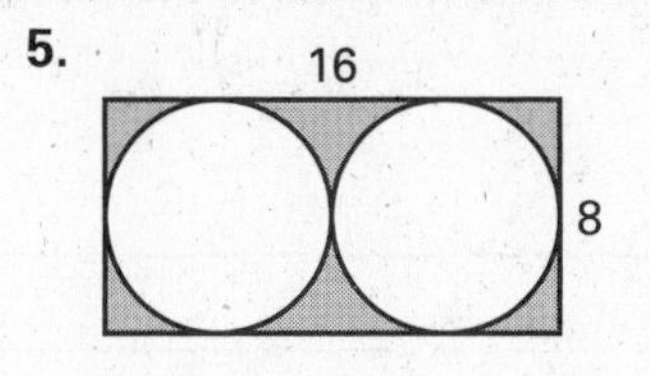

6.

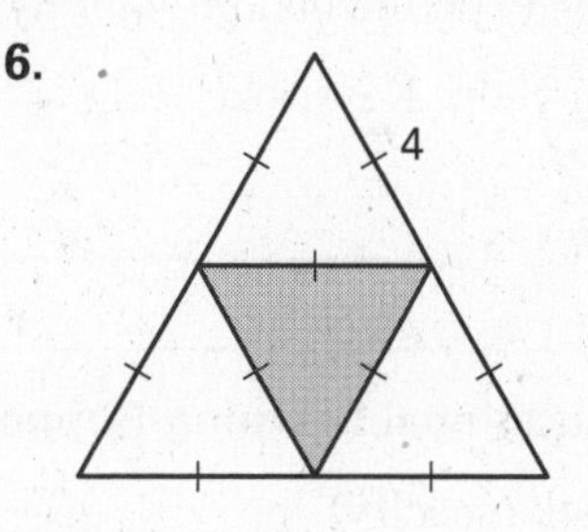

7.

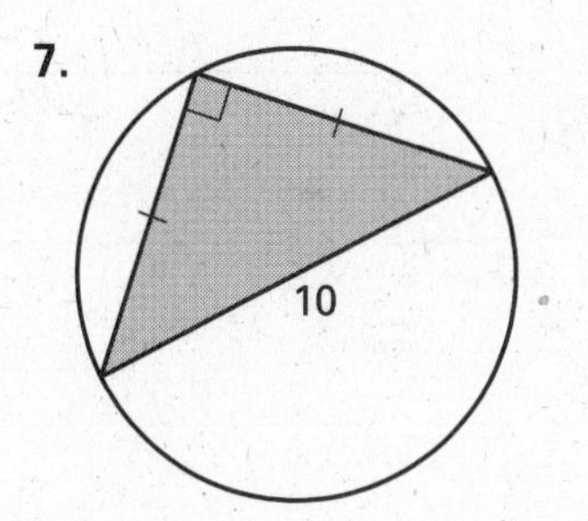

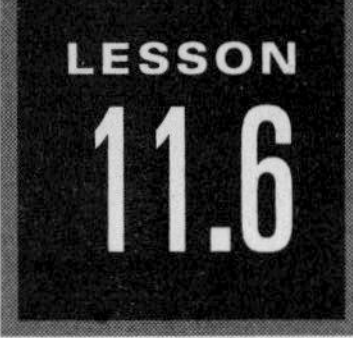

NAME ______________________________ DATE ____________

Quick Catch-Up for Absent Students

For use with pages 699–706

The items checked below were covered in class on (date missed) ____________

Lesson 11.6: Geometric Probability

____ **Goal 1:** Find a geometric probability. (p. 699)

Material Covered:

____ Student Help: Study Tip

____ Example 1: Finding a Geometric Probability

Vocabulary:

probability, p. 699 geometric probability, p. 699

____ **Goal 2:** Use geometric probability to solve real-life problems. (pp. 700–701)

Material Covered:

____ Example 2: Using Areas to Find a Geometric Probability

____ Example 3: Using a Segment to Find a Geometric Probability

____ Example 4: Finding a Geometric Probability

Activity 11.6: Investigating Experimental Probability (p. 706)

____ **Goal:** Find the experimental probability of an event using a graphing calculator simulation.

____ Student Help: Keystroke Help

____ Other (specify) ______________________________

____

Homework and Additional Learning Support

____ Textbook (specify) pp. 701–705

____

____ Internet: Extra Examples at www.mcdougallittell.com

____ *Reteaching with Practice* worksheet (specify exercises) ____________

____ *Personal Student Tutor* for Lesson 11.6

NAME ______________________________ DATE __________

Real-Life Application: When Will I Ever Use This?

For use with pages 699–705

Carnival Game

A game at a local carnival involves tossing beanbags at the target shown below. A person wins a prize if the beanbag goes through the eyes, nose, or part of the mouth.

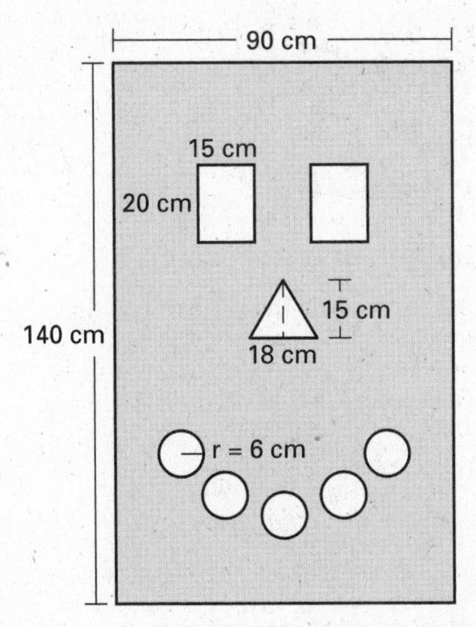

In Exercises 1–6, assume each person is throwing one beanbag. Round your results to three decimal places.

1. Find the probability that someone throws the beanbag through the left eye.
2. Find the probability that someone throws the beanbag through either eye.
3. Find the probability that someone throws the beanbag through the nose.
4. Find the probability that someone throws the beanbag through the mouth.
5. Find the probability that someone throws the beanbag through the circle at the far right of the mouth.
6. Find the probability that someone will win a prize.
7. Suppose that after the first night of the carnival, there are very few winners. A new target board is made with the lengths of the shapes on the target increased by two centimeters (The main board remains 140 cm by 90 cm.). The eyes are now 22 cm by 17 cm, the nose has a height of 17 cm and a base of 20 cm, and the circles for the mouth have a radius of 8 cm. Find the probability that someone will win a prize with the new target.

LESSON 11.6

NAME ________________________________ DATE ____________

Challenge: Skills and Applications

For use with pages 699–705

In Exercises 1–3, find the probability that a randomly chosen point in the figure lies in the shaded region. (Give your answer as a decimal rounded to the nearest hundredth.)

1.

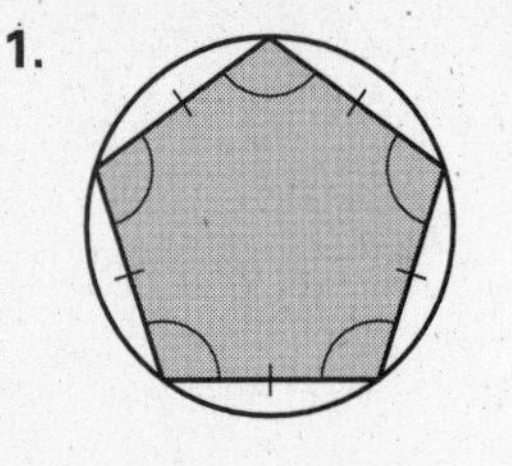

2.

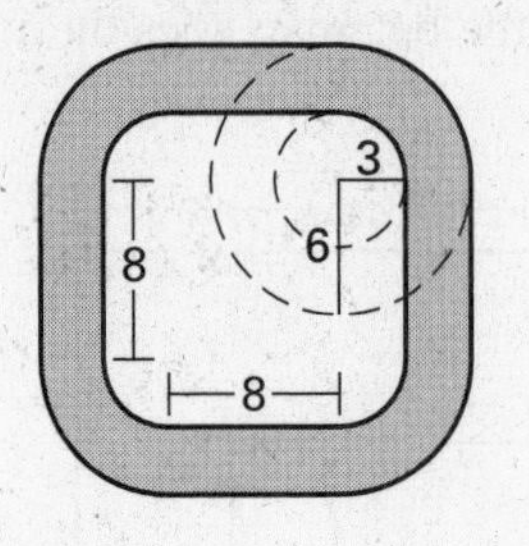

3.

In Exercises 4–6, refer to the diagram of a trapezoid.

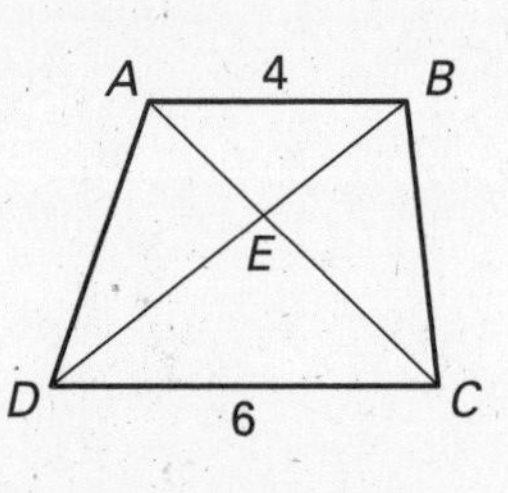

4. If a point on $\overline{BD}$ is chosen at random, what is the probability that the point is on $\overline{BE}$?

5. If a point in $\triangle ABC$ is chosen at random, what is the probability that the point is in $\triangle BCE$?

6. If a point in trapezoid $ABCD$ is chosen at random, what is the probability that the point is in $\triangle ABE$? in $\triangle CDE$? in $\triangle ADE$ or $\triangle BCE$?

In Exercises 7 and 8, use the following information.

At the 6th Street train station, the southbound trains arrive on the hour and at half past the hour, and the northbound trains arrive at 20 and 50 minutes after the hour. Each train remains at the station for 5 minutes before continuing its journey. Assume that the trains adhere rigidly to their schedule.

7. What is the probability that there is a train at the station when you arrive?

8. Suppose you want to travel south. If there is a northbound train at the station when you arrive, what is the probability that you will have to wait at least 8 minutes for the next southbound train to arrive?

NAME ______________________ DATE ________

Quiz 2

For use after Lessons 11.4–11.6

Find the indicated measure. *(Lesson 11.4)*

1. Radius **2.** Length of $\widehat{AB}$ **3.** Circumference

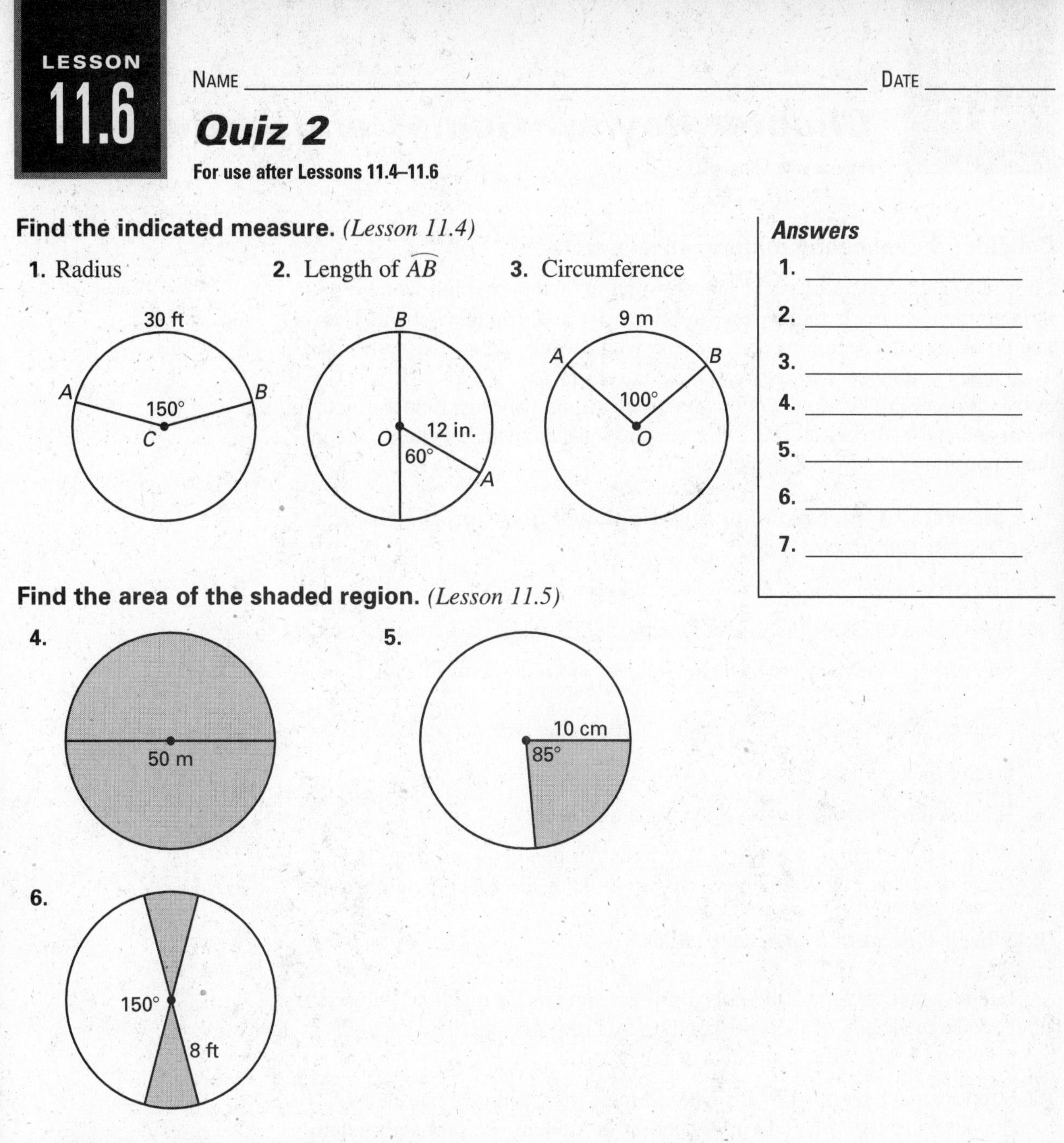

Answers

1. ________
2. ________
3. ________
4. ________
5. ________
6. ________
7. ________

Find the area of the shaded region. *(Lesson 11.5)*

4.

5.

6.

7. ***Targets*** A square target with 30 centimeter sides includes a rectangular region 2 centimeters by 3 centimeters. An arrow is shot and hits the target at random. Find the probability that the arrow hits the rectangle. *(Lesson 11.6)*

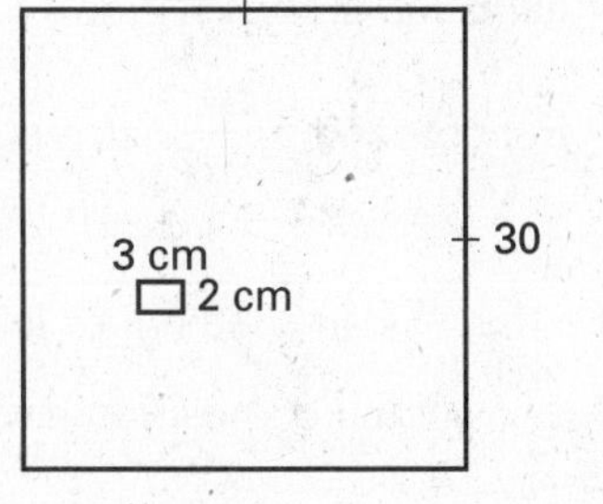

NAME ______________________ DATE __________

Chapter Review Games and Activities

For use after Chapter 11

Consider the following mathematical puzzle.

There are 100 numbered switches in a row, all in the off position. You go through the switches from Switch 1 to Switch 100 and flip all of them. Then you go back to the beginning and, starting with Switch 2, flip every other one. Then again you go back to the beginning and starting with Switch 3, you flip every third one. You continue in this fraction until on your last time through you begin and end with Switch 100. After you are finished, which switches are in the on position?

The answers to the following questions will give you the Switch numbers in the on position.

1. The circumference of a circle is 98π. What is its radius?
2. What is the length of the side of an equilateral triangle with an area of $4\sqrt{3}$?
3. Suppose that the measure of the intercepted arc of a sector of a circle is 60 degrees. Suppose also that the radius of the circle is $\sqrt{\frac{216}{\pi}}$. What is the area of the sector?
4. What is the probability of an event certain to occur?
5. What is the length of the apothem of a regular pentagon inscribed in a circle of radius 11? (Round your answer to the nearest whole number.)
6. What is the area of a circle with radius $\frac{5}{\sqrt{\pi}}$?
7. Suppose that polygon I and polygon II are similar. Each side of polygon II is twice the length of its corresponding side of polygon I. The area of polygon I is 16. What is the area of polygon II?
8. Suppose that region J contains region M and that region M has area 4. Suppose also that if a point in J is chosen at random, the probability that it is in region M is 0.25. What is the area of region J?
9. Suppose that arc $\widehat{AB}$ of a circle of radius $\frac{243}{\pi}$ has a measure of 60 degrees. What is the length?
10. What is the area of a square that has an apothem of length 5?
11. What do the numbered switches in the on position have in common?

NAME ______________________ DATE __________

Chapter Test A

For use after Chapter 11

In Exercises 1 and 2, use the figure at the right.

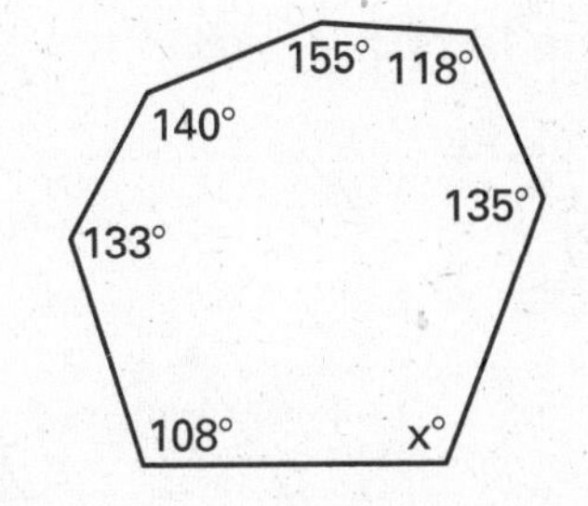

1. What is the value of x?

2. Find the sum of the measures of the exterior angles, one at each vertex.

In Exercises 3–6, you are given the number of sides of a regular polygon. Find the measure of each interior angle and each exterior angle.

3. 4 **4.** 8 **5.** 10 **6.** 15

In Exercises 7–10, you are given the measure of each interior angle of a regular *n*-gon. Find the value of *n*.

7. 120° **8.** 140° **9.** 160° **10.** 162°

In Exercises 11 and 12, find the perimeter and area of the regular polygon.

11.

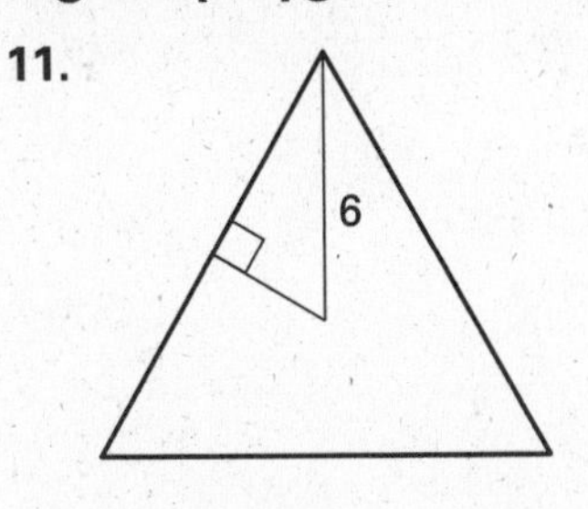

12.

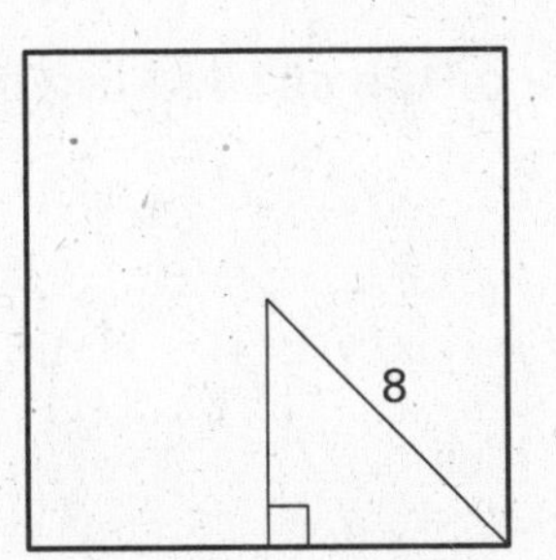

In Exercises 13 and 14, the polygons are similar. Find the ratio (large to small) of their perimeters and of their areas.

13.

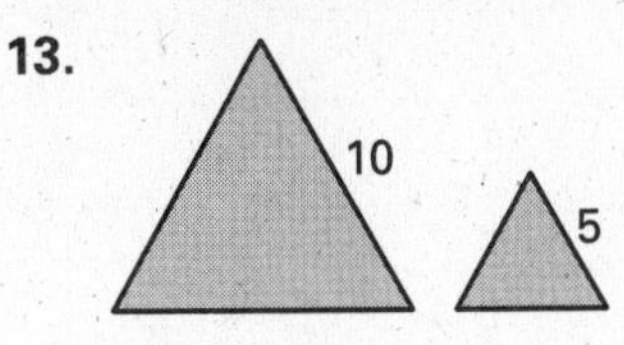

14.

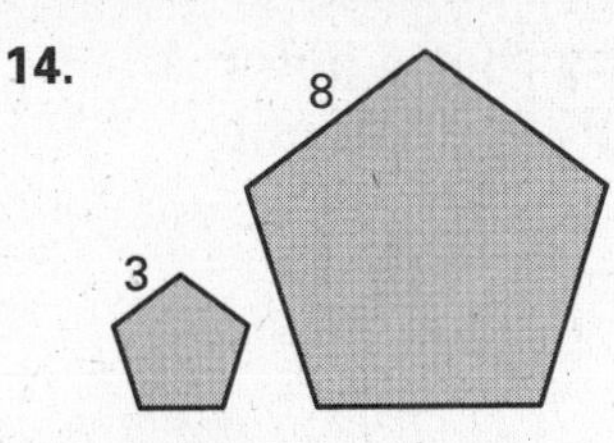

In Exercises 15 and 16, find the indicated measure.

15. Radius

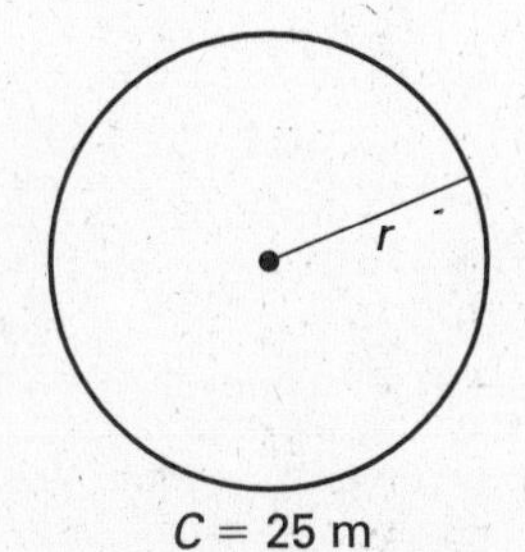

$C = 25$ m

16. Circumference

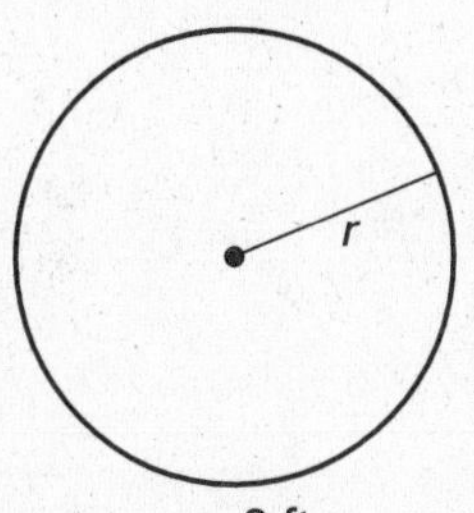

$r = 3$ ft

Answers

1. ________
2. ________
3. ________
4. ________
5. ________
6. ________
7. ________
8. ________
9. ________
10. ________
11. ________
12. ________
13. ________
14. ________
15. ________
16. ________

CHAPTER 11 CONTINUED

NAME ______________________ DATE __________

Chapter Test A

For use after Chapter 11

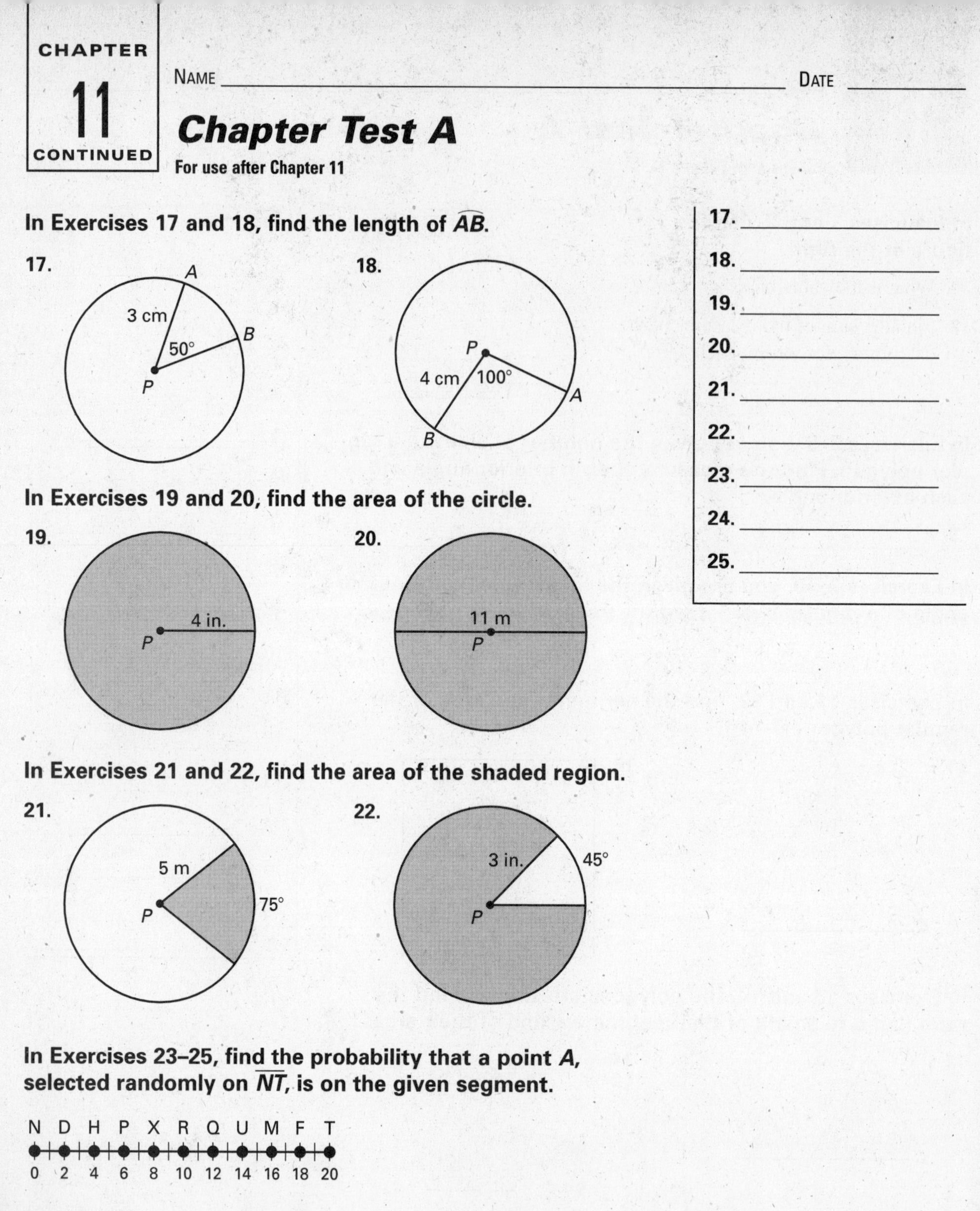

In Exercises 17 and 18, find the length of $\overset{\frown}{AB}$.

17.

18.

In Exercises 19 and 20, find the area of the circle.

19.

20.

In Exercises 21 and 22, find the area of the shaded region.

21.

22.

In Exercises 23–25, find the probability that a point *A*, selected randomly on $\overline{NT}$, is on the given segment.

23. $\overline{NX}$

24. $\overline{RU}$

25. $\overline{HF}$

17. ______
18. ______
19. ______
20. ______
21. ______
22. ______
23. ______
24. ______
25. ______

Review and Assess

NAME ______________________ DATE ____________

Chapter Test B

For use after Chapter 11

In Exercises 1 and 2, use the figure at the right.

1. What is the value of x?

2. Find the sum of the measures of the exterior angles, one at each vertex.

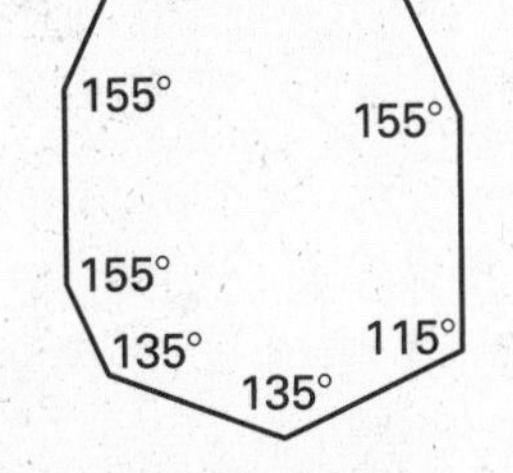

In Exercises 3–6, you are given the number of sides of a regular polygon. Find the measure of each interior angle and each exterior angle.

3. 5 **4.** 8 **5.** 9 **6.** 16

In Exercises 7–10, you are given the measure of each interior angle of a regular *n*-gon. Find the value of *n*.

7. 108° **8.** 157.5° **9.** 150° **10.** 170°

In Exercises 11 and 12, find the perimeter and area of the regular polygon.

11.

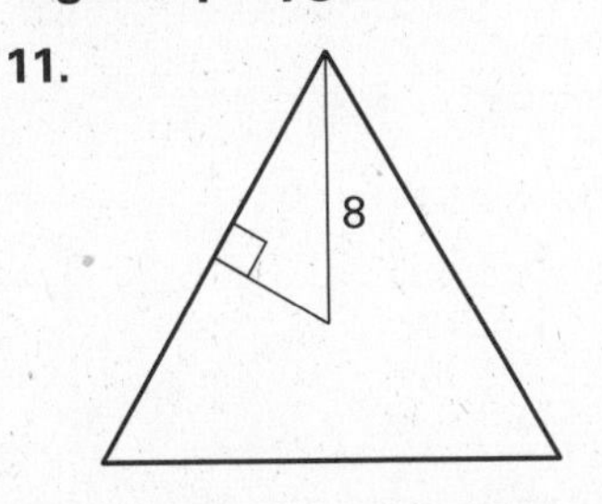

12.

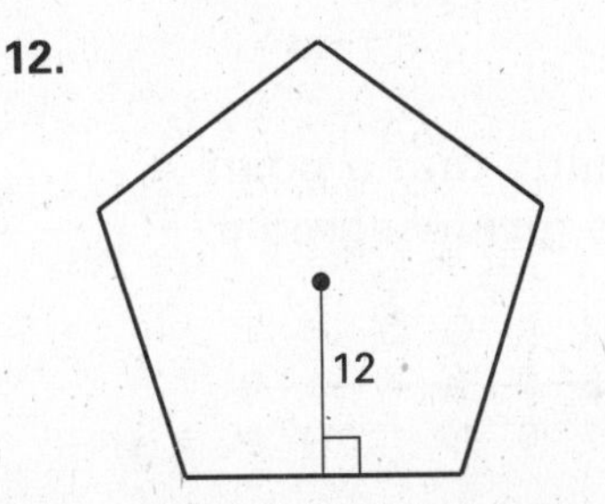

In Exercises 13 and 14, the polygons are similar. Find the ratio (large to small) of their perimeters and of their areas.

13. **14.**

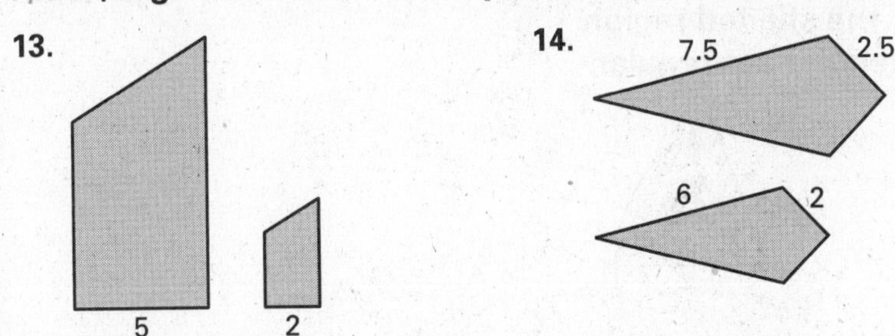

In Exercises 15 and 16, find the indicated measure.

15. Circumference

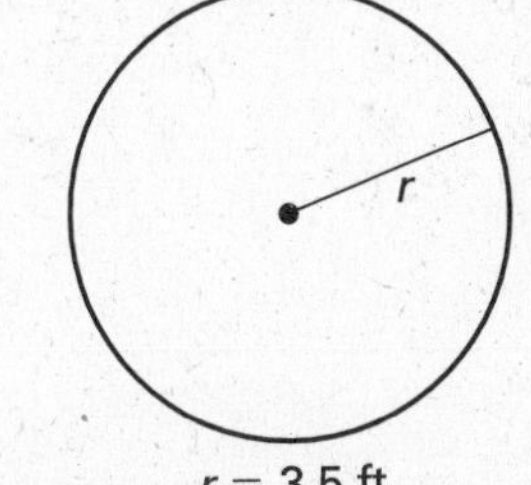

$r = 3.5$ ft

16. Radius

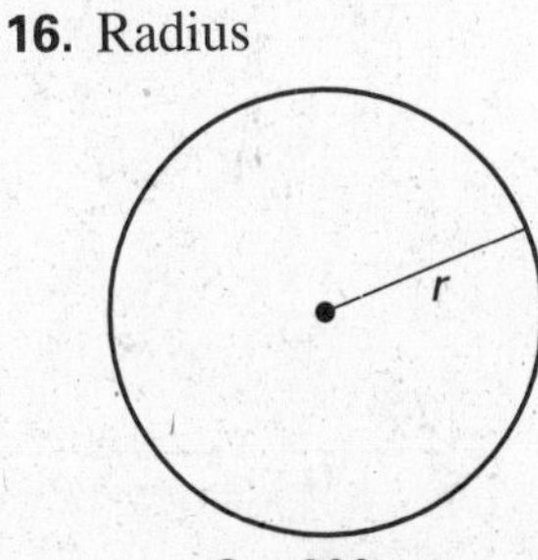

$C = 300$ cm

Answers

1. ____________
2. ____________
3. ____________
4. ____________
5. ____________
6. ____________
7. ____________
8. ____________
9. ____________
10. ____________
11. ____________
12. ____________
13. ____________
14. ____________
15. ____________
16. ____________

CHAPTER 11 CONTINUED

NAME ______________________ DATE __________

Chapter Test B

For use after Chapter 11

In Exercises 17 and 18, find the length of $\widehat{AB}$.

17. P, 6.5 m, 75°, B, A

18. A, 2 in., 150°, P, B

In Exercises 19 and 20, find the area of the shaded region.

19. P, 4.75 in.

20. 5.9 in., P, 80°

In Exercises 21–23, find the probability that a point *A*, selected randomly on $\overline{NT}$, is on the given segment.

21. $\overline{NR}$

22. $\overline{LQ}$

23. $\overline{DT}$

N D L U R Q B P T

0 2 4 6 8 10 12 14 16

In Exercises 24 and 25, find the probability that a point chosen at random in each figure is in the shaded region. Assume polygons that appear to be regular are regular.

24.

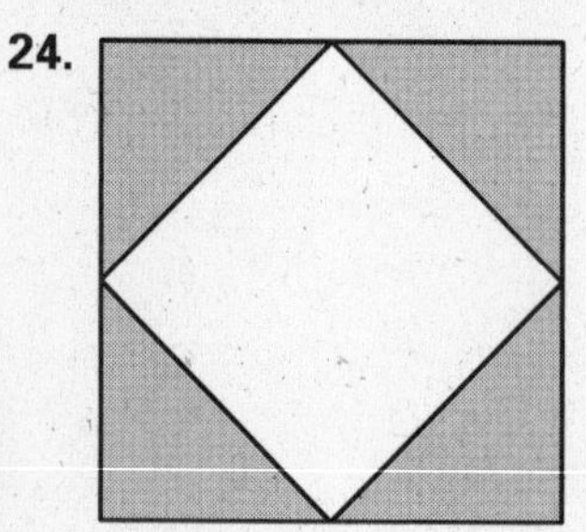

25.

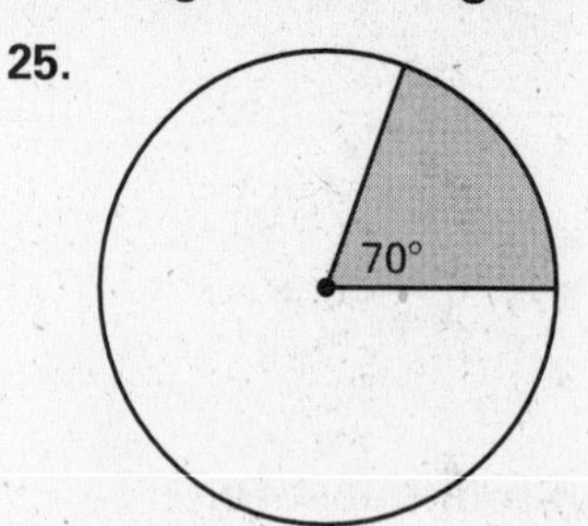

17. ________
18. ________
19. ________
20. ________
21. ________
22. ________
23. ________
24. ________
25. ________

Review and Assess

NAME ______________________ DATE __________

Chapter Test C

For use after Chapter 11

In Exercises 1 and 2, use the figure at the right.

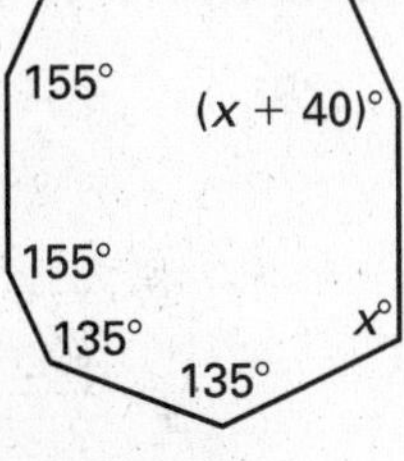

1. What is the value of x?
2. Find the sum of the measures of the exterior angles, one at each vertex.

In Exercises 3–6, you are given the number of sides of a regular polygon. Find the measure of each interior angle and each exterior angle.

3. 3 4. 6 5. 11 6. 21

In Exercises 7–10, you are given the measure of each interior angle of a regular *n*-gon. Find the value of *n*.

7. 90° 8. 135° 9. 150° 10. 170°

In Exercises 11 and 12, find the perimeter and area of the regular polygon.

11. 10

12. 10

In Exercises 13 and 14, the polygons are similar. Find the ratio (large to small) of their perimeters and of their areas.

13.

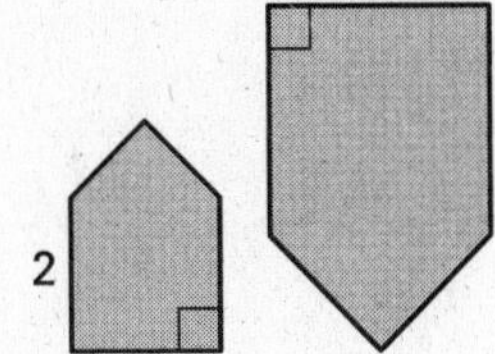

14.

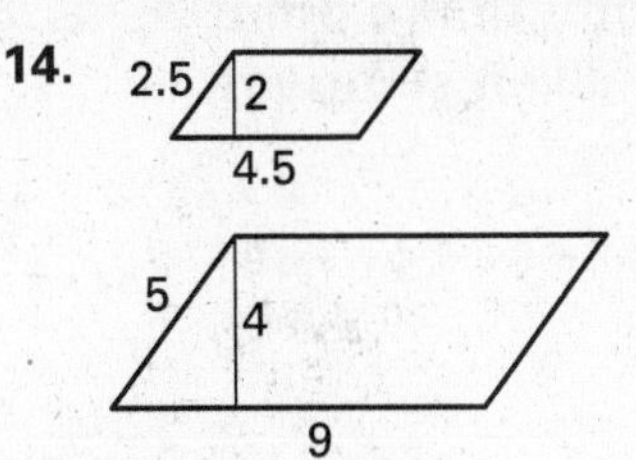

In Exercises 15 and 16, find the indicated measure.

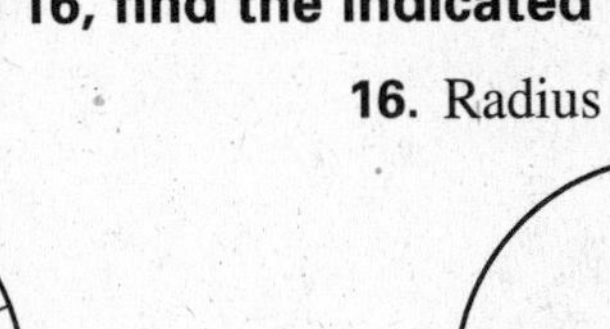

15. Circumference

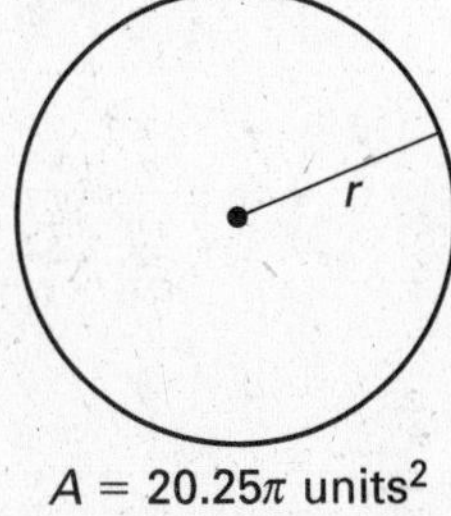

$A = 20.25\pi$ units^2

16. Radius

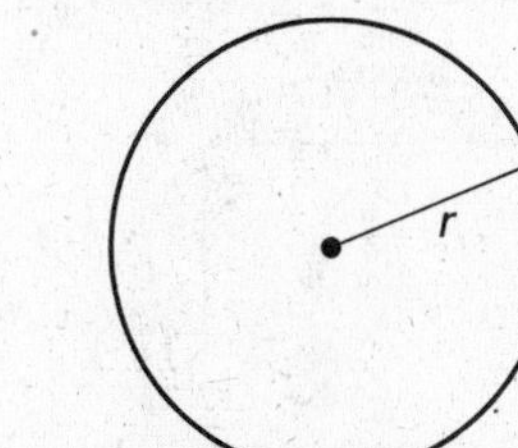

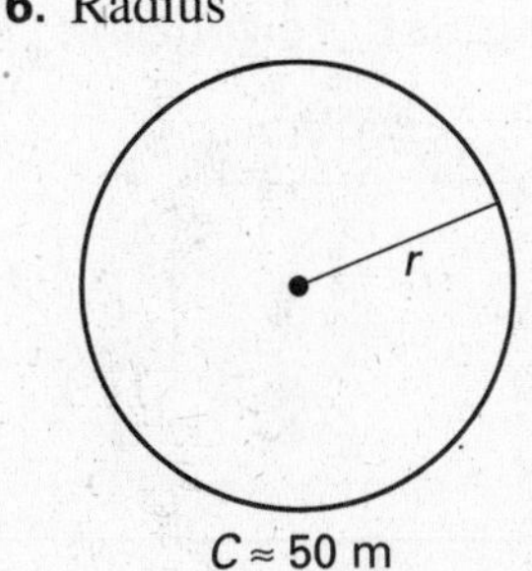

$C \approx 50$ m

Answers

1. ______
2. ______
3. ______
4. ______
5. ______
6. ______
7. ______
8. ______
9. ______
10. ______
11. ______
12. ______
13. ______
14. ______
15. ______
16. ______

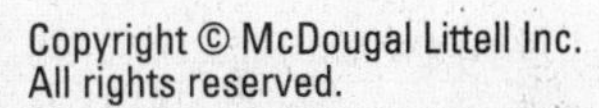

NAME ______________________ DATE __________

Chapter Test C

For use after Chapter 11

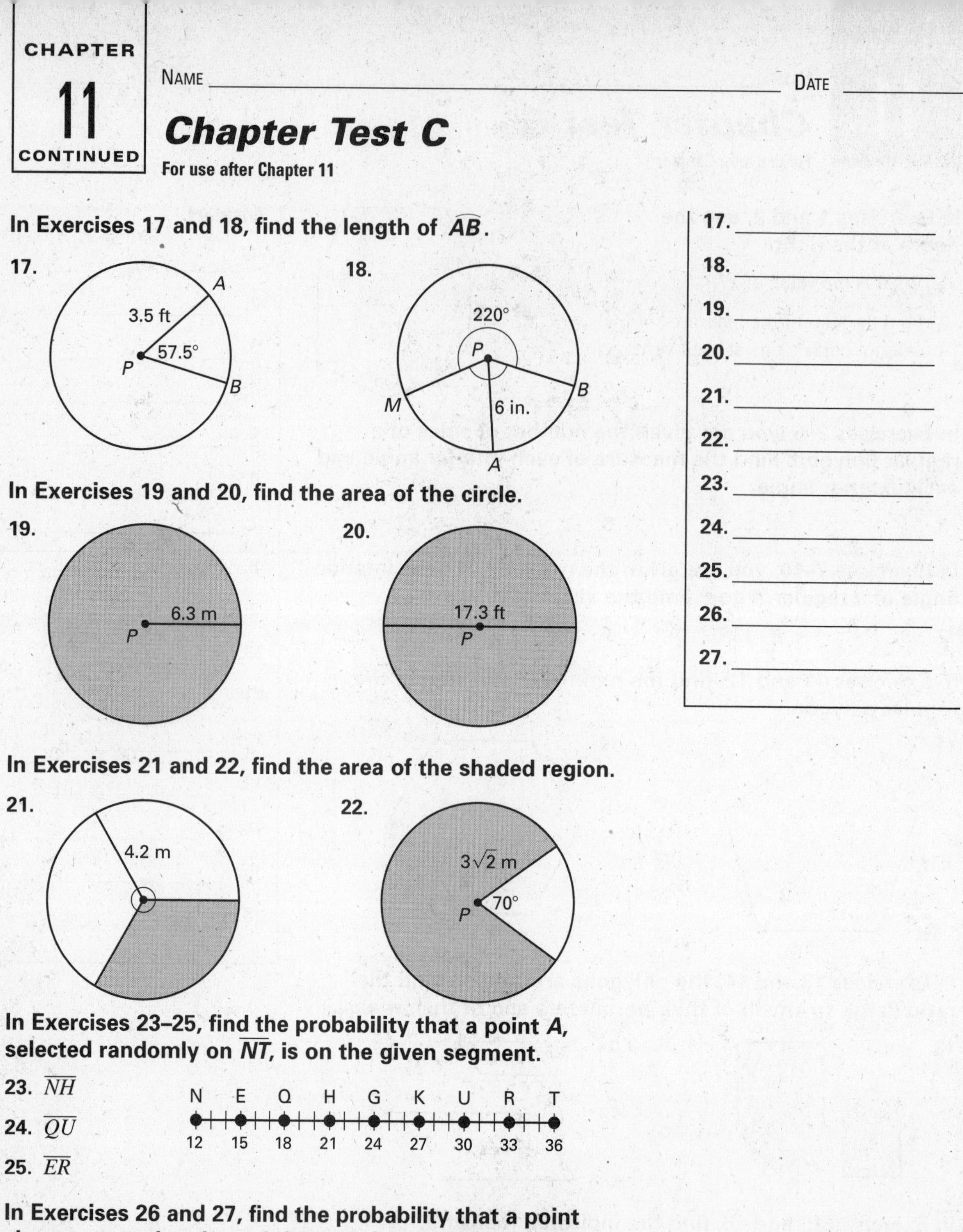

In Exercises 17 and 18, find the length of $\widehat{AB}$.

17.

18.

In Exercises 19 and 20, find the area of the circle.

19.

20.

In Exercises 21 and 22, find the area of the shaded region.

21.

22.

In Exercises 23–25, find the probability that a point A, selected randomly on $\overline{NT}$, is on the given segment.

23. $\overline{NH}$

24. $\overline{QU}$

25. $\overline{ER}$

17. ____________
18. ____________
19. ____________
20. ____________
21. ____________
22. ____________
23. ____________
24. ____________
25. ____________
26. ____________
27. ____________

In Exercises 26 and 27, find the probability that a point chosen at random in each figure is in the shaded region. Assume polygons that appear to be regular are regular.

26.

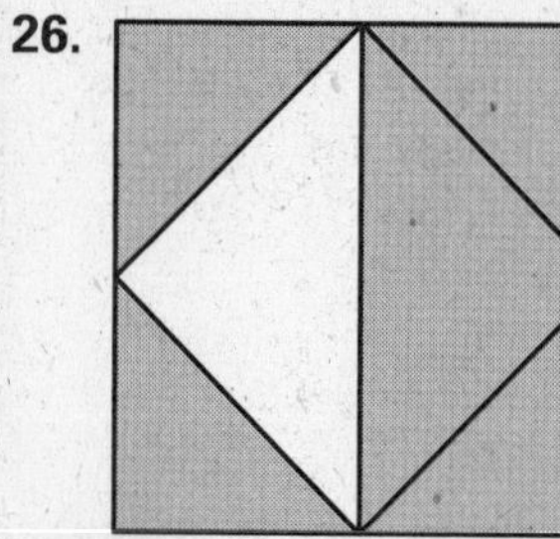

27.

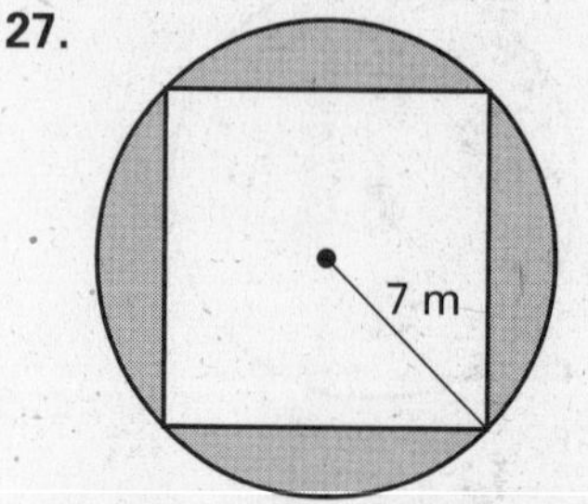

NAME ______________________________ DATE ____________

SAT/ACT Chapter Test

For use after Chapter 11

1. A regular polygon has an interior angle with a measure of 135°. How many sides does the polygon have?

Ⓐ 5 Ⓑ 6 Ⓒ 7
Ⓓ 8 Ⓔ 9

2. What is the value x?

Ⓐ 45
Ⓑ 50
Ⓒ 55
Ⓓ 60
Ⓔ 65

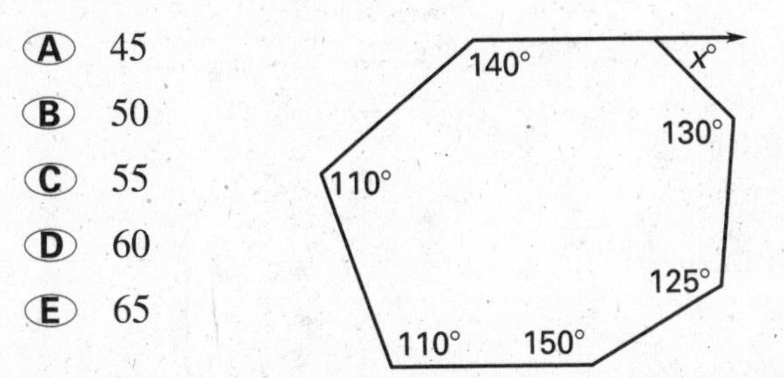

3. What is the length of $\widehat{AB}$?

Ⓐ about 2.45 m
Ⓑ about 28.27 m
Ⓒ about 9.77 m
Ⓓ about 68.42 m
Ⓔ about 34.21 m

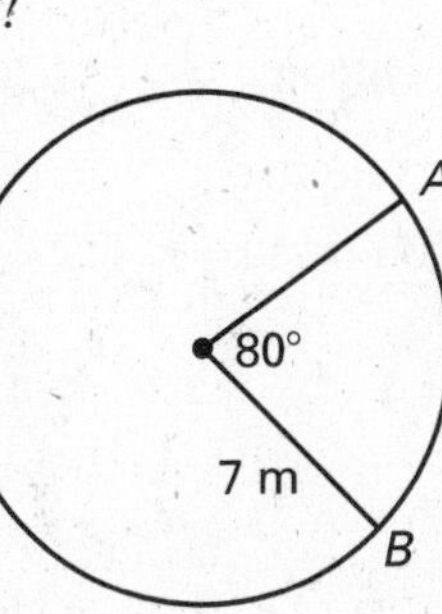

4. A regular hexagon has sides of length 8 cm. Another regular hexagon has sides of length 12 cm. Find the ratio of the area of the larger hexagon to the area of the smaller hexagon.

Ⓐ 4 : 9 Ⓑ 3 : 2
Ⓒ 9 : 4 Ⓓ 2 : 3
Ⓔ 12 : 8

5. What is the area of a regular pentagon if its apothem has a length of 4 feet and each side has a length of 5.8 feet?

Ⓐ 139.2 ft^2 Ⓑ 58 ft^2
Ⓒ 116 ft^2 Ⓓ 69.6 ft^2
Ⓔ 92.8 ft^2

6. What is the circumference of a circle with a radius of 4.2 inches?

Ⓐ 13.2 in. Ⓑ 6.6 in.
Ⓒ 8.4 in. Ⓓ 55.4 in.
Ⓔ 26.4 in.

7. Find the length of the apothem of a square if the area of the square is 64 square meters.

Ⓐ 2 m Ⓑ 4 m
Ⓒ $2\sqrt{2}$ m Ⓓ 8 m Ⓔ 6 m

8. What is the sum of the exterior angles in a convex 11-sided polygon?

Ⓐ 360° Ⓑ 540° Ⓒ 1620°
Ⓓ 180° Ⓔ 720°

9. What is the area of the shaded region in the diagram below?

Ⓐ about 19.72 m^2
Ⓑ about 14.59 m^2
Ⓒ about 12.33 m^2
Ⓓ about 10.27 m^2
Ⓔ about 8.11 m^2

3 cm

10. What is the measure of one exterior angle in a regular 8-sided polygon?

Ⓐ 45° Ⓑ 65° Ⓒ 90°
Ⓓ 120° Ⓔ 360°

Review and Assess

NAME ______________________ DATE ________

Alternative Assessment and Math Journal

For use after Chapter 11

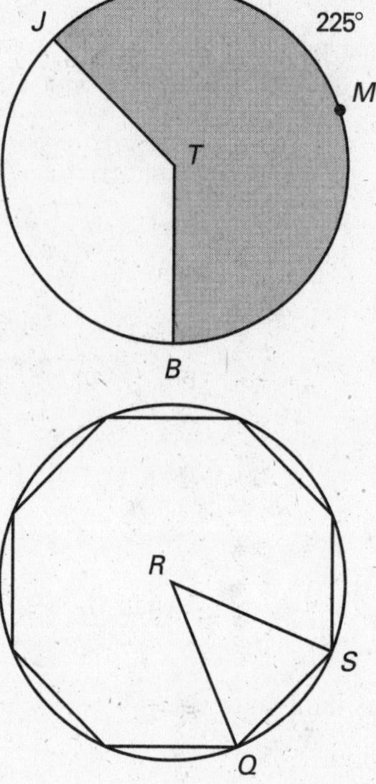

JOURNAL **1.** The length of $\overline{TJ}$ is 10 inches and $m\widehat{JMB}$ is 225°. (a.) Find the measure of angle *JTB*. Find the circumference of circle *T*. Find the arc length of $\widehat{JB}$. Find the area of circle *T*. Find the area of sector *JTB*. (b.) If the radius of circle *T* is doubled, will the area of sector *JTB* also double? Explain your reasoning by finding the area of sector *JTB* when the radius is doubled.

MULTI-STEP PROBLEM **2.** The regular octagon shown is inscribed in a circle with a radius of 5 inches. Each side of the octagon has a length of 3.8 inches.

a. Find the sum of the measures of the interior angles.

b. Find the measure of each interior angle.

c. Find the measure of each exterior angle.

d. Find the perimeter of the regular octagon.

e. Position point *T* on $\overline{SQ}$ so that the length of $\overline{RT}$ is the apothem of the octagon. Find the apothem *RT*.

f. Find the area of the regular octagon.

3. ***Critical Thinking***

a. Use the regular octagon from Exercise 2 to draw a similar octagon with a scale factor of 2 : 3. What is the perimeter of the larger octagon?

b. Find the probability that a randomly chosen point in the figure from Exercise 2 lies in the circle but not in the regular octagon.

4. ***Writing*** Write a paragraph proof and include a diagram.

GIVEN: $\triangle ABC$ is equilateral.

PROVE: Area of $\triangle ABC$ is $A = \frac{1}{4}\sqrt{3}s^2$.

Alternative Assessment Rubric

For use after Chapter 11

JOURNAL SOLUTION

1. Complete answers should include:

   a. $m\angle JTB = 135°$

   $C \approx 62.8$ inches

   Length of $\overset{\frown}{JB} \approx 23.6$ inches

   $A \approx 314$ square inches

   Area of sector $JTB \approx 117.75$ square inches

   b. No. When the radius is doubled to 20, the area of sector *JTB* is about 471 square inches. This is four times the original area.

MULTI-STEP PROBLEM SOLUTION

2. **a.** 1080°

   b. 135°

   c. 45°

   d. 30.4 inches

   e. 4.62 inches

   f. 70.2 square inches

3. **a.** 45.6 inches

   b. about 10.6%

4. Answers may vary.

MULTI-STEP PROBLEM RUBRIC

4 Students answer all questions correctly, showing their work. Students have correct diagrams. Students answer all parts of the proof correctly, showing work in a step-by-step manner.

3 Students complete the questions. Solutions may contain minor mathematical errors. Students show diagrams. Students complete the proof.

2 Students complete the questions. Several mathematical errors may occur. Students have insufficient diagrams. Students complete the proof, but some statements and reasons may not match or statements do not follow a logical order.

1 Students do not complete the questions. There is no work shown to support answers. Students have incomplete diagrams. Students' proof is incomplete. Final statement is not proven.

NAME ______________________ DATE 

Project: Gift Boxes

For use with Chapter 11

OBJECTIVE **Construct and find areas of special gift boxes.**

MATERIALS compass, metric ruler, scissors, heavy or card-stock paper

INVESTIGATION Use the pattern below for Exercises 1–4. Assume that the 14 minor arcs are congruent.

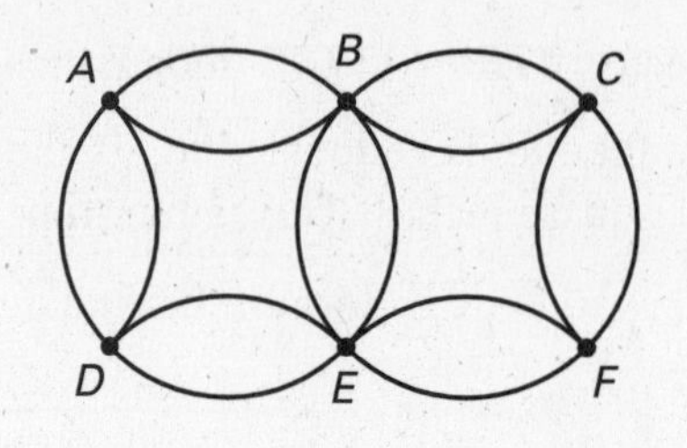

1. Draw a circle with radius 6 cm. Use this circle to construct a pattern like the one shown. (You may want to practice on regular paper before drawing the pattern on the heavy paper.) Record what you do at each step of your construction. (*Hint:* You need to first locate the endpoints of the arcs, and then use these points to locate the centers of the circles that form the arcs.)
2. Trace firmly along each arc with a pen to make creases.
3. Cut around the outer perimeter of the circles. Fold the sides of the box carefully along the arcs, with the pen marks on the inside. Your gift box is complete.
4. Construct two more gift boxes: one with radius 9 cm and one with radius 12 cm.
5. Organize a table of information for the three boxes as shown below.

	Radius	*Area*	$\frac{\text{radius}}{\text{radius of 1st box}}$	$\frac{\text{area}}{\text{area of 1st box}}$
1st box	6 cm		–	–
2nd box	9 cm			
3rd box	12 cm			

6. Find the area of the paper required for each box. Show your calculations. Record the results in your table.
7. Complete the last two columns of the table. Are the ratios of the areas between the figures what you would expect? Why or why not?

PRESENT YOUR RESULTS Write a report to go along with your boxes. Include the directions for your construction, the table of information for the three boxes, your calculations, and your discussion of the relationship between the areas.

Review and Assess

Project: Teacher's Notes

For use with Chapter 11

GOALS

- Analyze a drawing of circles and arcs.
- Construct circles and arcs.
- Use area formulas to solve a real-life problem.

MANAGING THE PROJECT Give students the chance to analyze the diagram and discuss it with each other. Then discuss it with the whole class. It is important for students to realize that the arcs are dividing the circumference of the circles in fourths. Students will need to carefully plan the placement of their construction on the paper. A class discussion about the construction might be helpful prior to the student work.

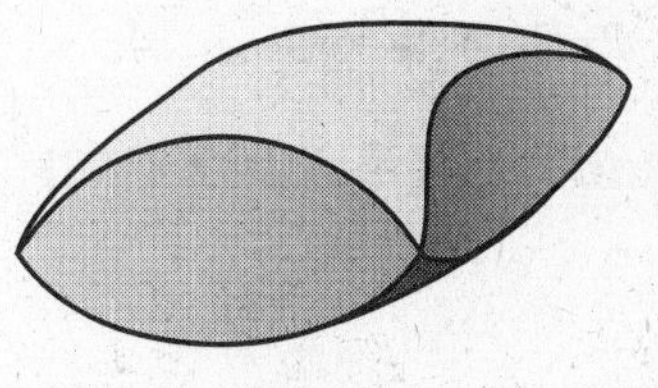

Students may need help determining the area of the pattern. Draw $\overline{BE}$ and show students that the area of the region bounded by $\widehat{BE}$ and $\overline{BE}$ is equal to the difference of the area of a 90° sector of the circle and the area of an isosceles right triangle. By doubling this area, students can find the area of the overlap region.

RUBRIC The following rubric can be used to assess student work.

4 The student neatly and correctly creates three different sized boxes and includes a clear and complete description of the construction used. The student presents an organized table of data including radii, area, and ratios, and calculations. The student understands and clearly states the relationship between the ratio of the radii and the ratio of the areas.

3 The student creates three different sized boxes and includes a description of the construction used. The description may be incomplete or somewhat unclear. The student presents an organized table of data including radii, area, and ratios, and calculations. Some calculations for the data may be missing or not explained. The student states the relationship between the ratio of the radii and the ratio of the areas.

2 The student creates two different sized boxes. The student presents a poorly organized table of data, which may be missing some entries. Some calculations for the data may contain errors, be missing, or lack explanation. The student attempts to describe the relationship between the ratio of the radii and the ratio of the areas, but fails to do so accurately.

1 The student attempts the construction but does not analyze the figure correctly. The student presents a poorly organized table of data, which may be missing some entries. Some calculations for the data may contain errors, be missing, or not be explained. Student omits the relationship between the ratio of the radii and the ratio of the areas.

NAME ______________________ DATE __________

Cumulative Review

For use after Chapters 1–11

Find the length and midpoint of $\overline{AB}$. (1.3, 1.5)

1. $A(6, -2), B(-4, 2)$

Solve the equation and state a reason for each step. (2.4)

2. $7x - 5 = 4x - 14$

Find the value of *x*. (3.3)

3.

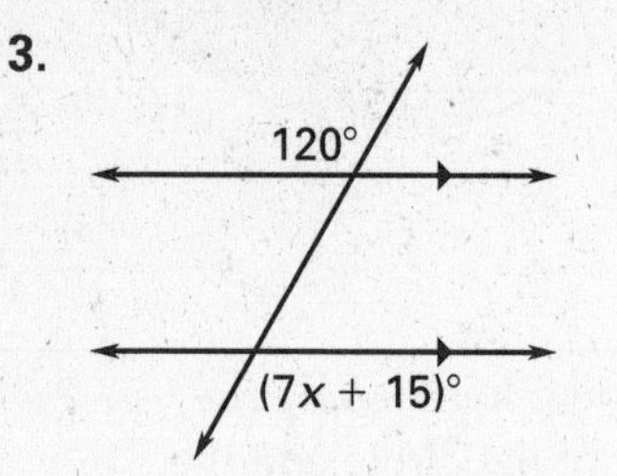

Write a two-column proof. (4.3, 4.4)

4. **Given:** $\overline{BC} \parallel \overline{DA}, \overline{BC} \cong \overline{DA}$

   Prove: $\triangle BAD \cong \triangle DCB$

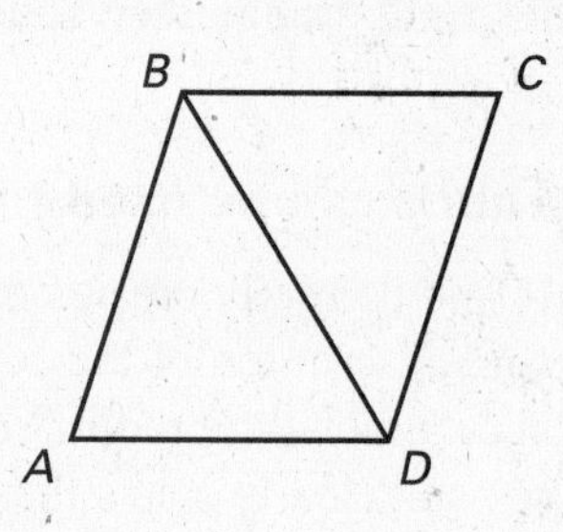

Find the value of *x*. (6.5)

5.

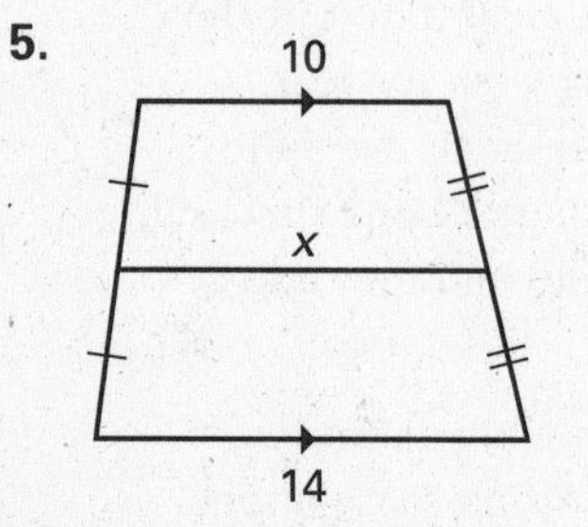

6.

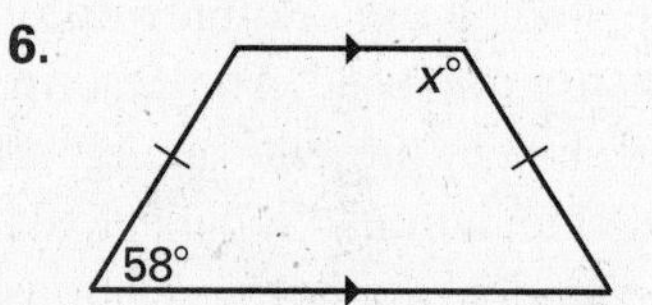

Find the coordinates of the reflection without using a coordinate plane. (7.2)

7. $S(2, 4)$ reflected in the x-axis.

8. $T(-2, 6)$ reflected in the y-axis.

The triangles are similar. Find the values of the variables. (8.6)

9.

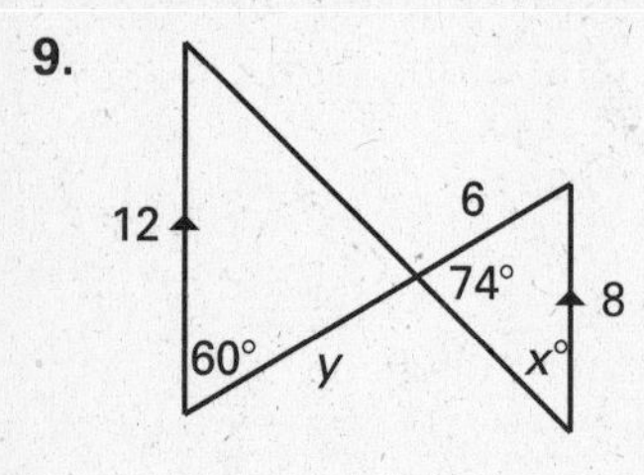

10.

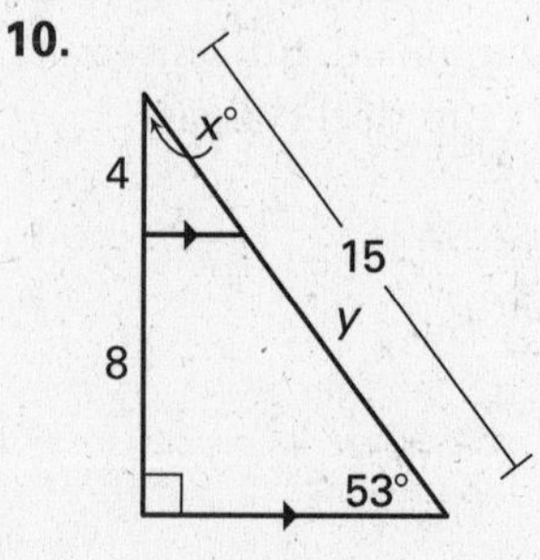

Review and Assess

NAME ______________________ DATE ____________

Cumulative Review

For use after Chapters 1–11

Use the diagram to find the measure of $\angle A$ and $\angle B$ to the nearest degree. (9.6)

11.

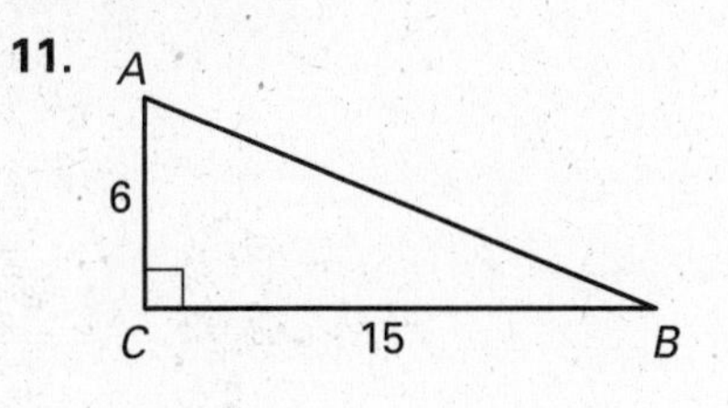

12.

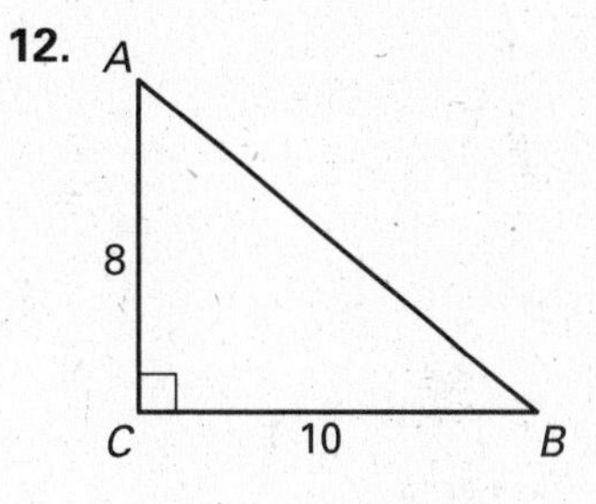

Find the value of x. (11.1)

13.

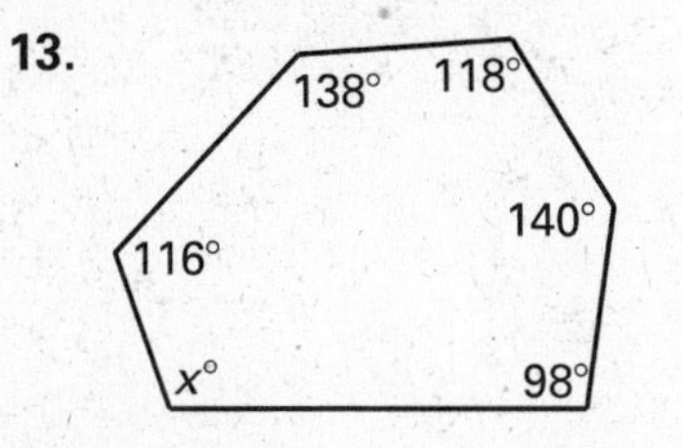

14.

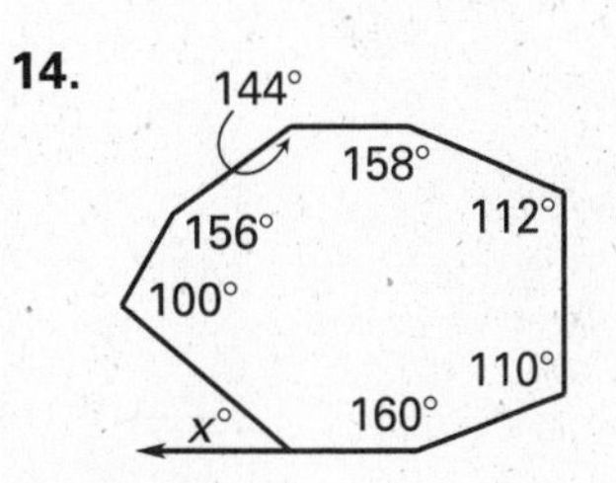

Find the perimeter and area of the regular polygon. (11.2)

15. **16.** **17.**

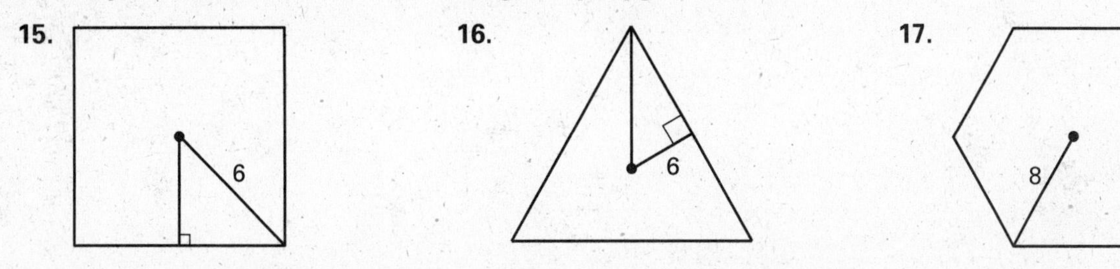

The polygons below are similar. Find the ratio (large to small) of their perimeters and of their areas. (11.3)

18.

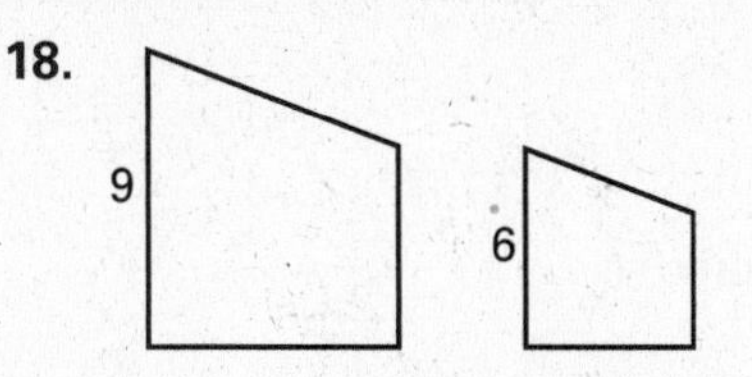

19.

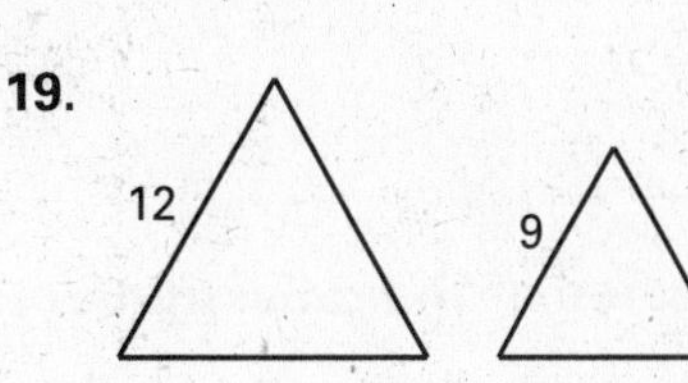

Find the area of the shaded region. Use $\pi \approx 3.14$ and round your answer to two decimal places. (11.5)

20. **21.**

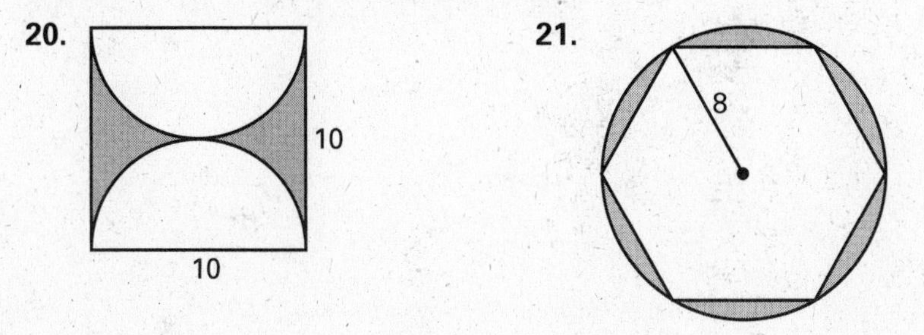

22.

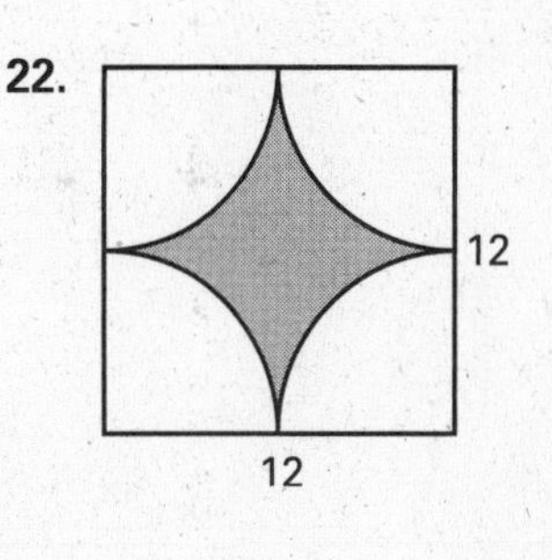

Review and Assess

ANSWERS

Chapter Support

Parent Guide

11.1: pentagon **11.2:** about 65 in.^{2}; 89 tiles **11.3:** about \$20.23 **11.4:** about 176 cm **11.5:** about 339 ft^2 **11.6:** about 9.4%

Prerequisite Skills Review

1. $m\angle B = 60°, m\angle 1 = 90°, m\angle 2 = 120°, m\angle 3 = 150°$ **2.** $m\angle B = 40°, m\angle 1 = 110°, m\angle 2 = 140°, m\angle 3 = 110°$

3. $m\angle B = 30°, m\angle 1 = 50°, m\angle 2 = 150°, m\angle 3 = 160°$ **4.** 14 cm^2 **5.** 22.5 in.^{2}

6. 24.5 units^2 **7.** $\frac{3}{5}$ **8.** $\frac{5}{3}$ **9.** $\frac{3}{5}$ **10.** $\frac{5}{3}$ **11.** $\frac{5}{3}$ **12.** $\frac{5}{3}$

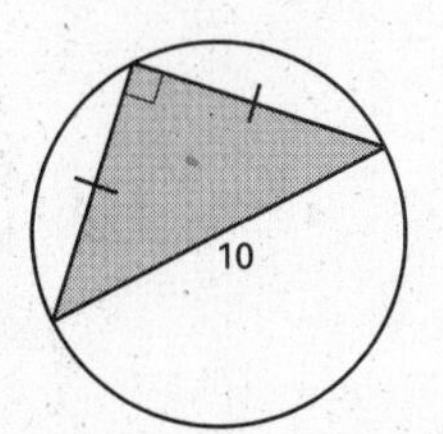

Strategies for Reading Mathematics

1. It is the first letter of the word area; they are the first letters of the words length and width; they are the first letters of the words base and height.

2. what it means and the part of the figure whose measure it represents **3.** $b = 2$ m and $h = 3$ m; yes; the units are the same. **4.** $\ell = 4$ ft and $w = 10$ in.; no; the units are not the same.

Lesson 11.1

Warm-Up Exercises

1. 40° **2.** 48° **3.** 60°

4. equilateral or equiangular triangle

Daily Homework Quiz

1–3. Check students' drawings.

1. 2 lines on opp. sides of *f*, each parallel to *f* and 1 in. from *f* and all the points between the lines **2.** circle *p* with radius 1 cm and all points outside the circle

3. line with slope $\frac{1}{2}$ passing through (2, 0)

Lesson Opener

Allow 10 minutes.

1. equilateral triangle; 60° **2.** square; 90°

3. regular pentagon; 108° **4.** regular hexagon; 120° **5.** regular heptagon; $\approx 129°$ **6.** regular octagon; 135° **7.** regular nonagon; 140°

8. regular decagon; 144°

9. regular dodecagon; 150°

Practice A

1. 4, 4 **2.** 6, 6 **3.** 10, 10 **4.** 5, 5 **5.** 720° **6.** 1080° **7.** 1800° **8.** 2340° **9.** 96 **10.** 85 **11.** 97 **12.** 117 **13.** 120 **14.** 135 **15.** 4 **16.** 5 **17.** 8 **18.** 10 **19.** 360° **20.** 360° **21.** 360° **22.** 360° **23.** 4 **24.** 6 **25.** 8 **26.** 12 **27.** 12 **28.** 15

Practice B

1. 540° **2.** 1260° **3.** 1980° **4.** 2880° **5.** 39 **6.** 116 **7.** 101 **8.** 124 **9.** 108 **10.** 135 **11.** 5 **12.** 7 **13.** 12 **14.** 20 **15.** 360° **16.** 360° **17.** 360° **18.** 360° **19.** 5 **20.** 8 **21.** 18 **22.** 24 **23.** 7 **24.** 13 **25.** 8 **26.** 10

Practice C

1. 900° **2.** 1440° **3.** 2520° **4.** 3960° **5.** 146 **6.** 120 **7.** 82 **8.** 8 **9.** 15 **10.** 20 **11.** 50 **12.** 9 **13.** 10 **14.** 48 **15.** 180 **16.** 15 **17.** 26 **18.** No; the polygon would have 14.4 sides which is not possible. **19.** Yes; the polygon would have 18 sides. **20.** Yes; the polygon would have 24 sides. **21.** Yes; the polygon would have 30 sides. **22.** always **23.** never **24.** always **25.** sometimes **26.** always **27.** sometimes

Reteaching with Practice

1. 150 **2.** 145 **3.** 33 **4.** 15 **5.** 52 **6.** 15

Lesson 11.1 *continued*

Cooperative Learning

1. 360° **2.** 360° **3.** yes

Interdisciplinary Application

1. 140° **2.** 40° **3.** 30.6 ft **4.** about 71.5 ft^2

Challenge: Skills and Applications

1. 6 **2.** 8 **3.** 10 **4.** 127° **5.** 147° **6.** 36°

7. 20° **8.** 30° **9.** 72°

10. not necessarily regular;

Sample answer:

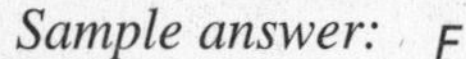

11. must be regular; *Sample answer:* First, note that $\angle H$, $\angle I$, and $\angle J$ must all be interior angles. (That is, the angles all have the same "orientation." For example, sides $\overline{GH}$, $\overline{HI}$, and $\overline{IJ}$ cannot be arranged so that $\overline{GH}$ and $\overline{IJ}$ are parallel, for if they were, G and I would be more than two side-lengths apart and it would be impossible to complete the pentagon.) Now draw $\overline{GI}$ and $\overline{IK}$. By the SAS Congruence Postulate, $\triangle GHI \cong \triangle KJI$. This gives $\angle HGI \cong \angle JKI$, and also $\overline{GI} = \overline{KI}$, which by the Base Angles Theorem gives $\angle HGK \cong \angle GKJ$. Using algebra and the fact that the sum of the measures of the interior angles of a pentagon is $3 \cdot 180°$, we find that these angles measure 108°. Since all sides are congruent and all angles are congruent, $GHIJK$ is a regular pentagon.

12. must be regular; *Sample answer:* Assume there are at least 4 sides, because an equilateral triangle is known to be regular. Let A, B, C, and D be any four vertices. Since $\overline{AB}$, $\overline{BC}$, and $\overline{CD}$ are congruent chords in the same circle, $\overset{\frown}{AB} \cong \overset{\frown}{BC} \cong \overset{\frown}{CD}$ and, using the Arc Addition Postulate, $\overset{\frown}{ABC} \cong \overset{\frown}{BCD}$. Therefore, $\angle ADC \cong \angle BAD$. In this manner, it can be shown that all of the angles are congruent, so P is a regular polygon.

13. not necessarily regular; *Sample answer:* A nonsquare rectangle can be inscribed in a circle.

Lesson 11.2

Warm-Up Exercises

1. 10 cm **2.** 60° **3.** 30° **4.** $5\sqrt{3}$ cm

Daily Homework Quiz

1. 2520° **2.** 124 **3.** 20 **4.** 60

Lesson Opener

Allow 10 minutes.

1–3.

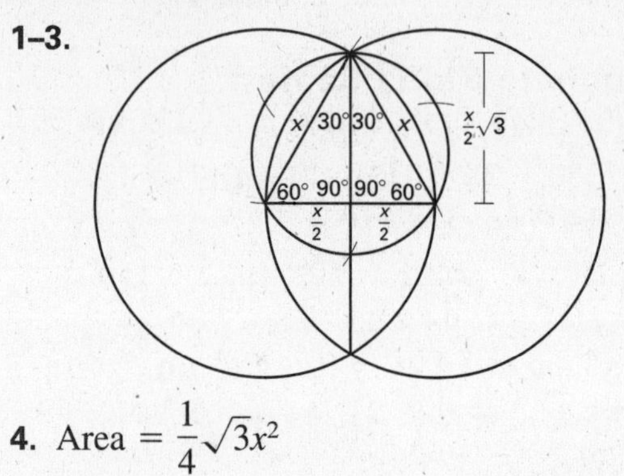

4. Area $= \frac{1}{4}\sqrt{3}x^2$

Technology Activity

1. Answers will vary. *Sample answer:*

$A = 6\left(\frac{1}{4}\sqrt{3}s^2\right) = \frac{3}{2}\sqrt{3}s^2$, where s is the length of a side of the equilateral triangle.

2. $24\sqrt{3} \approx 41.57$ square units **3.** Answers will vary. *Sample answer:* Find the length of one of the sides of the hexagon and use the formula from the answer to Exercise 1 (because this length is also the side length of one of the six equilateral triangles that form the regular hexagon).

Practice A

1. G **2.** 4 **3.** $\angle AGB$ **4.** $\overline{GH}$ **5.** 6

6. 42.5 cm^2 **7.** $72\sqrt{3}$ in.^{2} **8.** 86.4 cm^2

9. $\sqrt{3}$ units^2 **10.** $9\sqrt{3}$ units^2

11. $\frac{25\sqrt{3}}{4}$ units^2 **12.** 64 units^2

13. $24\sqrt{3}$ units^2 **14.** about 58.1 units^2

15. 24 units; 41.57 units^2 **16.** $32\sqrt{2}$ units; 128 units^2 **17.** 53.02 units; 212.08 units^2

Lesson 11.2 *continued*

Practice B

1. $16\sqrt{3}$ units^2 2. $\frac{169\sqrt{3}}{4}$ units^2 3. $8\sqrt{3}$ units^2
4. 45° 5. 36° 6. 20° 7. 15°
8. $24\sqrt{3}$ units; $48\sqrt{3}$ units^2 9. $24\sqrt{2}$ units; 72 units^2 10. about 72.65 units; about 363.27 units^2 11. 72 units; about 374.12 units^2
12. about 73.48 units; about 407.29 units^2
13. about 74.16 units; about 423.2 units^2
14. false 15. true 16. false 17. false
18. 36 in.^{2} 19. about 173.82 in.^{2} 20. $25.33

Practice C

1. $25\sqrt{3}$ units^2 2. $\frac{6.25\sqrt{3}}{4}$ units^2
3. $48\sqrt{3}$ units^2 4. 30° 5. 24° 6. 14.4°
7. 11.25° 8. 60 units; about 247.75 units^2
9. about 59.65 units; about 268.41 units^2
10. about 43.26 units; about 144.01 units^2
11. about 292.28 cm^2 12. about 194.86 in.^{2}
13. 19.845 cm^2 14. about 6.2 cm 15. always
16. always 17. sometimes 18. never
19. about 6.93 in.^{2} 20. about 41.57 in.^{2}
21. $22.\overline{2}$ % 22. $9.36

Reteaching with Practice

1. 21.2 cm^2 2. 36.7 in.^{2} 3. 60.3 m^2
4. 21.4 square units 5. 73.44 ft, 407.13 ft^2

Real-Life Application

1. 4605 ft 2. 72°; isosceles 3. about 633.8 ft
4. about 1,459,324.5 ft 5. about $\frac{1}{9}$

Challenge: Skills and Applications

1. *OM*: apothem, *OP*: radius, *PQ*: side length
2. $x = \frac{180°}{n}$ 3. $a = r\cos\left(\frac{180}{n}\right)^\circ$
4. $s = 2r\sin\left(\frac{180}{n}\right)^\circ$ 5. $a = \frac{s}{2\tan\left(\frac{180}{n}\right)^\circ}$
6. $\text{area} = \frac{ns^2}{4\tan\left(\frac{180}{n}\right)^\circ}$
7. $\text{area} = na^2\tan\left(\frac{180}{n}\right)^\circ$
8. $\text{area} = nr^2\sin\left(\frac{180}{n}\right)^\circ\cos\left(\frac{180}{n}\right)^\circ$
9. 2, 2.83, 3.11, 3.13, 3.14; 3.14 or π
10. a. *Sample answer:* Draw $\overline{PA}$, $\overline{PB}$, $\overline{PC}$, $\overline{PD}$, and $\overline{PE}$. Let $s = AB$. Then the area of *ABCDE*
$= \text{area}(\triangle ABP) + \text{area}(\triangle BCP) + \text{area}(\triangle CDP) + \text{area}(\triangle DEP) + \text{area}(\triangle EAP)$
$= \frac{1}{2}s \cdot PV + \frac{1}{2}s \cdot PW + \frac{1}{2}s \cdot PX + \frac{1}{2}s \cdot PY + \frac{1}{2}s \cdot PZ = \frac{1}{2}s \cdot (PV + PW + PX + PY + PZ)$
So, $PV + PW + PX + PY + PZ = \frac{2}{s} \cdot$ (area of *ABCDE*), which does not depend on how *P* is chosen inside the pentagon. b. 17.2

Lesson 11.3

Warm-Up Exercises

1. 1:3 2. 33 3. 9 4. 1:3

Daily Homework Quiz

1. $\frac{81\sqrt{3}}{4} \approx 35.1$ square units 2. 36°
3. 392 square units 4. $24\sqrt{3} \approx 41.6$ in.^{2}

Lesson Opener

Allow 15 minutes.

1. *Sample answer:* Formula in A3 gives the number of the next triangle; Formulas in B3, C3, and D3 give integer multiples of a 3-4-5 triangle; Formula in E2 gives the perimeter of the triangle; Formula in F2 gives the area of each triangle; Formulas in G2, H2, and H3 compare the shortest sides, perimeter, and area respectively to that of the 3-4-5 triangle.

Lesson 11.3 *continued*

Ratios

	A	B	C	D	E	F	G
1	Triangle	a	b	c	Perimeter	Area	Ratio of sides
2	1	3	4	5	12	6	1
3	2	6	8	10	24	24	2
4	3	9	12	15	36	54	3
5	4	12	16	20	48	96	4
6	5	15	20	25	60	150	5
7	6	18	24	30	72	216	6
8	7	21	28	35	84	294	7
9	8	24	32	40	96	384	8
10	9	27	36	45	108	486	9
11	10	30	40	50	120	600	10
12	11	33	44	55	132	726	11
13	12	36	48	60	144	864	12
14	13	39	52	65	156	1014	13
15	14	42	56	70	168	1176	14
16	15	45	60	75	180	1350	15
17	16	48	64	80	192	1536	16
18	17	51	68	85	204	1734	17
19	18	54	72	90	216	1944	18
20	19	57	76	95	228	2166	19
21	20	60	80	100	240	2400	20

Ratios

H	I
Ratio of perimeters	Ratio of areas
1	1
2	4
3	9
4	16
5	25
6	36
7	49
8	64
9	81
10	100
11	121
12	144
13	169
14	196
15	225
16	256
17	289
18	324
19	361
20	400

2. *Sample answer:* The ratio of the perimeters of two similar triangles is the *same* as the ratio of their corresponding sides. The ratio of the areas of two similar triangles is the *square* of the ratio of their corresponding sides.

Practice A

1. 3:7; 9:49 **2.** 2:1, 4:1 **3.** 2:3; 4:9 **4.** 3:2; 9:4 **5.** sometimes **6.** always **7.** always **8.** sometimes **9.** 25:64 **10.** 4:3 **11.** 4:5 **12.** 100 cm^2 **13.** 3.8 cm by 2 cm **14.** 3.8 m by 2 m **15.** 7.6 cm^2:7.6 m^2 = 1 cm^2:1 m^2

Practice B

1. 3:5; 9:25 **2.** 7:5; 49:25 **3.** 7:10; 49:100 **4.** 9:7; 81:49 **5.** 9:49 **6.** 2:$\sqrt{3}$ **7.** $\sqrt{5}$:2$\sqrt{2}$ **8.** 138.24 cm^2 **9.** 40 mm × 60 mm **10.** 20 m × 30 m **11.** 2400 mm^2; 600 m^2 **12.** 150$\sqrt{3}$ mm^2; $\frac{75}{2}\sqrt{3}$ m^2 **13.** 200 mm^2; 50 m^2 **14.** 1500 mm^2; 375 m^2 **15.** ≈ 260.05 m^2

Practice C

1. $\sqrt{12}$:5; 12:25 **2.** 14:11; 196:121 **3.** 3:5; 9:25 **4.** 5:4; 25:16 **5.** 64:9 **6.** $\sqrt{21}$:$\sqrt{10}$ **7.** 4$\sqrt{2}$:3$\sqrt{5}$ **8.** 8$\sqrt{2}$ cm by 12$\sqrt{2}$ cm **9.** 9:4 **10.** 4:9 **11. a.** $\frac{49}{64}$ in.^{2}; $248\frac{1}{16}$ ft^2 **b.** $\frac{21}{64}$ in.^{2}; $106\frac{5}{16}$ ft^2

Reteaching with Practice

1. 2:5, 4:25 **2.** 1:5, 1:25 **3.** 3:5, 9:25 **4.** 3:5, 9:25 **5.** 1.7 ft **6.** 84 sec **7.** \$1886.40 **8.** 9.2 ft^2 **9.** No, in real life, the cow would be 132 inches long, or 11 feet long. This is not a realistic length for a cow.

Interdisciplinary Application

1. 1:2; 1:4 **2.** 1:16; yes **3.** \$1.92; \$3.00

Math and History Application

1. Since $AC = 1$, $AB = \frac{1}{2}$. Furthermore, $m\angle ABD = \frac{360°}{6} = 60°$. Hence, $\triangle ABD$ is equilateral, and $AD = \frac{1}{2}$. The perimeter of the inscribed hexagon is therefore $6\left(\frac{1}{2}\right) = 3$.

2. Since $m\angle ADB = 60°$, $m\angle EDF = 30°$. Since $\triangle EDF$ is a right triangle, $m\angle EFD = 60°$. Since the sides of a 30° - 60° - 90° triangle are in the ratio x, $\sqrt{3}x$, and $2x$, we have that

$$EF = \frac{1}{\sqrt{3}} \cdot ED = \frac{1}{\sqrt{3}}\left(\frac{1}{4}\right) = \frac{\sqrt{3}}{12}$$ and that

$$FD = 2\left(\frac{\sqrt{3}}{12}\right) = \frac{\sqrt{3}}{6}.$$ Hence,

$$FG = 2 \cdot FD = 2\left(\frac{\sqrt{3}}{6}\right) = \frac{\sqrt{3}}{3}.$$

Lesson 11.3 *continued*

3. The perimeter of the circumscribed hexagon would be $6 \cdot FG = 6\left(\frac{\sqrt{3}}{3}\right) = 2\sqrt{3}$.

4. Since

perimeter of inscribed polygon < circumference of circle < perimeter of circumscribed polygon

we have that $3 < \pi < 2\sqrt{3}$.

Challenge: Skills and Applications

1. a. 75 **b.** 6 **c.** 10 **d.** 192

2. a. $\frac{y}{x}, \frac{z}{x}$ **b.** $\frac{Ay^2}{x^2}, \frac{Az^2}{x^2}$

c. *Sample answer:* Since $A + \frac{Ay^2}{x^2} = \frac{Az^2}{x^2}$ and $A \neq 0$, we can multiply both sides of the equation by $\frac{x^2}{A}$, giving $x^2 + y^2 = z^2$.

3. 6 **4.** 16 **5.** 10 **6.** 9

Quiz 1

1. 3960° **2.** 15° **3.** $100\sqrt{3} \approx 173.2 \text{ in.}^2$
4. $\approx 745.58 \text{ cm}^2$ **5.** $\frac{3}{1}; \frac{9}{1}$ **6.** $\frac{12}{5}; \frac{144}{25}$ **7.** \$288

Lesson 11.4

Warm-Up Exercises

1. 41° **2.** 110° **3.** 250° **4.** 319°

Daily Homework Quiz

1. $\frac{5}{7}, \frac{25}{49}$ **2.** $\triangle ACD$ is similar to $\triangle AEB$. Area of $\triangle AEB$ is $83\frac{1}{3}$. **3.** The price is very reasonable; it is almost half the cost of 4 times the cost of the smaller tablecloth.

Lesson Opener

Allow 15 minutes.

1. Check drawings.

2.

Diameter of pizza (in inches)	*Length of outer crust of one piece (in sixteenths of an inch)*
4	$3\frac{2}{16}$
11	$5\frac{12}{16}$
15	$5\frac{14}{16}$
17	$4\frac{7}{16}$

3. Check drawings. It should include 4 segments of lengths shown in the table for Exercise 2. *Sample answer:* A reasonable length for the outer crust of a piece of pizza is 3 to 6 inches. If a larger pizza were cut into the same number of pieces as a smaller pizza, the outer crust would be too long and the piece of pizza would be awkward to eat.

Practice A

1. F **2.** D **3.** C **4.** B **5.** A **6.** E
7. 12π in. **8.** 16π cm **9.** 13 in. **10.** 20 cm
11. 3.5π in. **12.** 4π ft **13.** 3π cm
14. a. about 18.85 ft/2 rev.
b. about 94.25 ft/10 rev.
15. a. about 14.66 ft/2 rev.
b. about 73.30 ft/10 rev.
16. a. about 10.47 ft/2 rev.
b. about 52.36 ft/10 rev.
17. a. about 29.32 ft/2 rev.
b. about 146.61 ft/10 rev.
18. about 23.56 cm **19.** 90 in. **20.** 36 in.

Practice B

1. 8.4π cm **2.** 7π in. **3.** about 2.13 in.
4. about 9.87 ft **5.** about 7.85 in.
6. about 30.79 cm **7.** about 19.63 in.
8. about 7.94 in. **9.** 64 cm **10.** about 8.11 cm
11. about 45.32 units **12.** about 36.57 units
13. about 62.83 units **14.** about 28.57
15. about 35.6 in.

Practice C

1. 11.4π cm **2.** about 14.96 in.
3. about 35.94 in. **4.** about 34.94 cm
5. about 50.82 cm **6.** about 6.16 in.

Lesson 11.4 *continued*

7. $18 + 6\pi$ cm **8.** 28π in. **9.** $16 + 4\pi$ in. **10.** 1414 cm **11.** $(40 + 10\pi) \approx 71.42$ m; $(40 + 18\pi) \approx 96.55$ m **12.** about 39.10 in. **13.** $8 + 6\pi \approx 26.85$ in.

Reteaching with Practice

1. 106.81 cm **2.** 43.98 in. **3.** 2.23 yd **4.** 3.82 ft **5.** 4.19 **6.** 3.49 **7.** 5.87 **8.** 26.13 **9.** 66.21 **10.** 103.1°

Real-Life Application

1. $r_3 = 39.09$ m, $r_4 = 40.33$ m, $r_5 = 41.57$ m, $r_6 = 42.81$ m, $r_7 = 44.05$ m, $r_8 = 45.29$ m

2. Lane 1: $\approx$ 400.03 m, Lane 2: $\approx$ 407.82 m, Lane 3: $\approx$ 415.61 m, Lane 4: $\approx$ 423.40 m, Lane 5: $\approx$ 431.19 m, Lane 6: $\approx$ 438.98 m, Lane 7: $\approx$ 446.77 m, Lane 8: $\approx$ 454.57 m

3. about 8 meters

4. Lane 1: $\approx$ 115.01 m, Lane 2: $\approx$ 118.91 m, Lane 3: $\approx$ 122.80 m, Lane 4: $\approx$ 126.70 m, Lane 5: $\approx$ 130.60 m, Lane 6: $\approx$ 134.49 m, Lane 7: $\approx$ 138.39 m, Lane 8: $\approx$ 142.28 m

5. You and your two friends will not run equal distances since there is an 8 meter difference in the lengths of each lane.

Challenge: Skills and Applications

1. 24π **2.** $15 + 5\pi$ **3.** $45 + 25\pi$ **4.** 2.01 **5.** 4.18 **6.** 4.22 **7. a.** 7.2°, Alternate Interior Angles Theorem **b.** 25,000 mi; 3979 mi **c.** *Sample answer:* Eratosthenes' estimate was within 20 miles, or 0.5%, of the actual radius.

Lesson 11.5

Warm-Up Exercises

1. 144 in.^{2} **2.** about 10.8 cm^2 **3.** 14.5 ft^2 **4.** about 665 yd^2

Daily Homework Quiz

1. 18.84 in. **2.** 40π m **3.** 1.74 cm **4.** 75 **5.** 18π

Lesson Opener

Allow 10 minutes.

1. *Sample answer:* The area of the half circle is half the area of the full circle and twice the area of the quarter circle.

2.

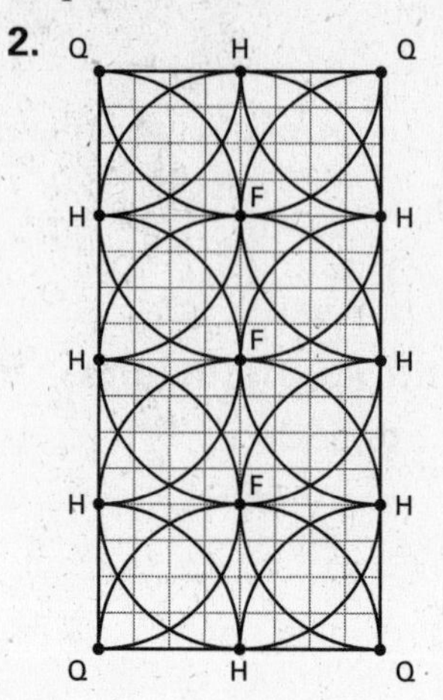

4 quarter circles are used at the corners, 8 half circles are used along the sides, and 3 full circles are used in the middle.

3.

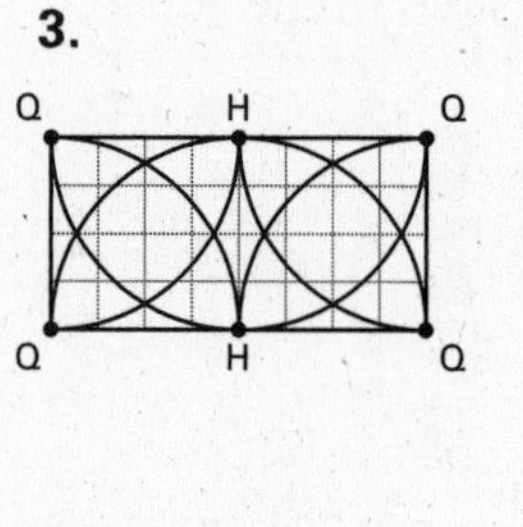

4.

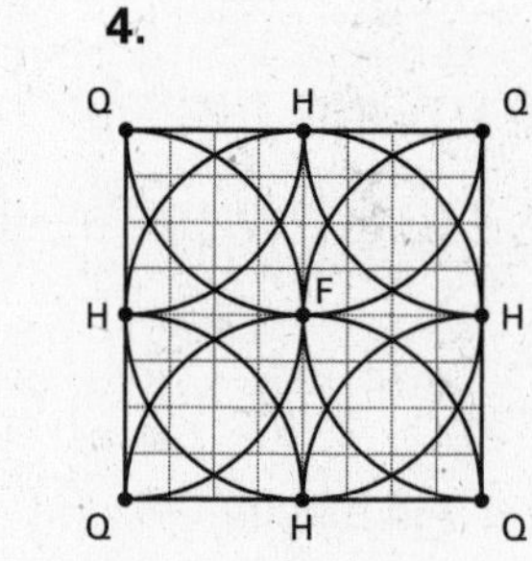

Technology Activity

1. Area of circle A = 2 (Area of circle D + Area of Circle E)

$$\pi r^2 = 2\left[\pi\left(\frac{1}{2}r\right)^2 + \pi\left(\frac{1}{2}r\right)^2\right]$$
$$= 2\left(\pi \cdot \frac{1}{4}r^2 + \pi \cdot \frac{1}{4}r^2\right)$$
$$= 2\left(\pi \cdot \frac{1}{2}r^2\right)$$
$$= \pi r^2$$

2. *Sample answer:* If the radius of $\odot A$ is r, then the circumference of $\odot A$ is $2\pi r$. The radius of each of the smaller circles is $\frac{1}{2}$ the length of the radius of $\odot A$. So, the circumference of each smaller circle is $2\pi\left(\frac{1}{2}r\right)$ or πr. When the circumference of $\odot D$ is added to the circumference of $\odot E$, the result is $2\pi r$, which equals the circumference of $\odot A$.

Lesson 11.5 *continued*

Practice A

1. D **2.** B **3.** E **4.** A **5.** C
6. about 201.06 ft^2 **7.** about 0.79 yd^2
8. about 28.27 cm^2 **9.** about 52.36 $in.^2$
10. about 126.1 m^2 **11.** about 603.19 ft^2
12. about 37.7 $in.^2$ **13.** about 21.46 cm^2
14. about 73.06 ft^2 **15.** 5 cm **16.** 12 in.
17. about 3.91 ft

Practice B

1. about 0.44 $in.^2$ **2.** about 63.62 cm^2
3. about 33.51 $in.^2$ **4.** about 21.21 $in.^2$
5. about 29.45 $in.^2$ **6.** about 221.74 $in.^2$
7. about 122.52 cm^2 **8.** about 29.48 $in.^2$
9. about 31.42 cm^2 **10.** about 106.81 cm^2
11. about 73.96 ft^2 **12.** about 20.55 ft^2
13.

arc	30°	60°	90°	120°	150°	180°
area of sector	4.2	8.4	12.6	16.8	20.9	25.1

Practice C

1. about 43.01 cm^2 **2.** about 103.87 $in.^2$
3. about 157.28 mm^2 **4.** about 150.56 $in.^2$
5. about 58.08 cm^2 **6.** about 28.36 $in.^2$
7. about 31.42 m^2 **8.** about 31.32 m^2
9. about 70.69 $in.^2$ **10.** about 110.01 $in.^2$
11. about 219.75 cm^2 **12.** about 12.57 cm^2
13. 15.5 ft **14.** 1.75 m **15.** c. 16-inch pizza
16. They are all the same amount of waste.

Reteaching with Practice

1. 3019.07 cm^2 **2.** 7.82 in. **3.** 5.17 ft
4. 7.07 $in.^2$ **5.** 131.14 ft^2 **6.** 34.91 m^2

Interdisciplinary Application

1. about 52,333 mi^2
2. about 17,662.5 mi^2 **3.** about 5652 mi^2

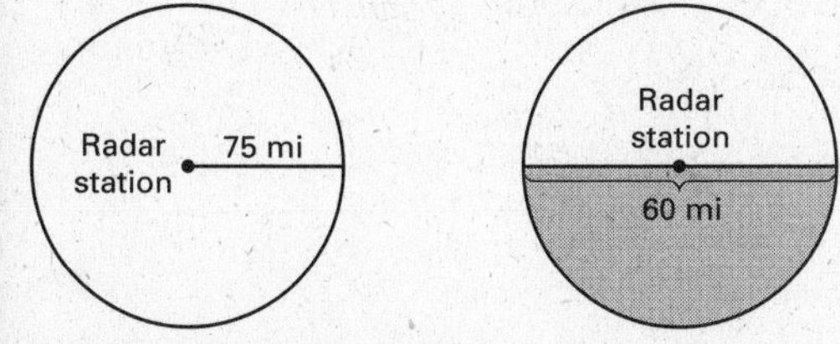

Challenge: Skills and Applications

1. $5\pi - 8$ **2.** $16\pi + 24\sqrt{3}$

3. $48\pi + 72\sqrt{3}$ **4. a.** $\frac{11}{2}r^2 + \frac{9}{4}\pi r^2$

b. 1.54 cm **5.** *Sample answer:* Let c be the diameter of the large circle. Since a diagonal of the rectangle is a diameter of the circle, $a^2 + b^2 = c^2$. Then:

(area of shaded region) = (total area) − (area of large circle)

= (area of rectangle) + (area of 4 semicircle) − (area of large circle)

$$= ab + \pi\left(\frac{a}{2}\right)^2 + \pi\left(\frac{b}{2}\right)^2 - \pi\left(\frac{c}{2}\right)^2$$

$$= ab + \frac{\pi}{4}(a^2 + b^2 - c^2)$$

$$= ab$$

6. $\sqrt{2}, \sqrt{3}, 2$ **7.** 18 $in.^2$

Lesson 11.6

Warm-Up Exercises

1. 100π cm^2, or about 314 cm^2
2. $150\sqrt{3}$ cm^2, or about 260 cm^2
3. $100\pi - 150\sqrt{3}$ cm^2, or about 54 cm^2

Daily Homework Quiz

1. 81π cm^2 **2.** about 8.7 ft^2
3. $1800 - 450\pi$ $in.^2$ **4.** 10 m

Lesson Opener

Allow 15 minutes.

Teacher's Note: You might want to have each group choose a different shape.

1. Check gameboards. **2.** Answers will vary.
3. Answers will vary. **4. a.** 25% **b.** 25%
c. 25% **d.** 25% **e.** ≈ 24%; Answers will vary. Students may comment that in their game they were using whatever throwing skill they had to try to make the penny land inside the shape, so it was not true that every point on the gameboard was an equally likely landing point. If they were skillful, they may find that their percents in Exercise 3 are significantly greater than the geometric probabilities found in Exercise 4.

Lesson 11.6 *continued*

Practice A

1. 20% **2.** 20% **3.** about 33% **4.** about 67% **5.** about 17% **6.** about 33% **7.** about 33% **8.** about 83% **9.** 50% **10.** 21.5% **11.** 36.3% **12.** 75% **13.** 64.3% **14.** 25% **15.** 25% **16.** $38.\overline{8}\%$ **17.** $8.\overline{3}\%$ **18.** $19.\overline{4}\%$ **19.** $8.\overline{3}\%$ **20.** $33.\overline{3}\%$

Practice B

1. about 17% **2.** 25% **3.** about 33% **4.** 75% **5.** about 8% **6.** 50% **7.** 50% **8.** about 92% **9.** 80% **10.** 21.5% **11.** $66.\overline{6}\%$ **12.** 9.3% **13.** 74.3% **14.** 75% **15.** about 7.32% **16.** $11.\overline{1}\%$ **17.** $16.\overline{6}\%$

Practice C

1. about 17% **2.** 25% **3.** about 42% **4.** about 67% **5.** about 67% **6.** about 83% **7.** 62.5% **8.** 21.5% **9.** 17.3% **10.** 18.75% **11.** 75% **12.** 50% **13.** 40% **14.** about 31.25% **15. a.** 16% **b.** 20% **c.** 28% **d.** 36%

Reteaching with Practice

1. 0.5 **2.** 0.375 **3.** 0.625 **4.** 0.25 **5.** 0.2146 **6.** 0.25 **7.** 0.3183

Real-Life Application

1. 0.024 **2.** 0.048 **3.** 0.011 **4.** 0.045 **5.** 0.009 **6.** 0.103 **7.** 0.153

Challenge: Skills and Applications

1. 0.76 **2.** 0.49 **3.** 0.58 **4.** $\frac{2}{5}$ **5.** $\frac{3}{5}$ **6.** $\frac{4}{25}, \frac{9}{25}, \frac{12}{25}$ **7.** $\frac{1}{3}$ **8.** $\frac{2}{5}$

Quiz 2

1. 11.46 ft **2.** 25.1 m **3.** 32.4 m **4.** 1963.5 m^2 **5.** 74.2 cm^2 **6.** 33.5 ft^2 **7.** $\approx$ 0.7%

Review and Assessment

Review Games and Activities

1. 49 **2.** 4 **3.** 36 **4.** 1 **5.** 9 **6.** 25 **7.** 64 **8.** 16 **9.** 81 **10.** 100 **11.** Only the switches with a number that is a perfect square will end up in the on position.

Test A

1. 111 **2.** 360° **3.** 90°, 90° **4.** 135°, 45° **5.** 144°, 36° **6.** 156°, 24° **7.** 6 **8.** 9 **9.** 18 **10.** 20 **11.** $18\sqrt{3}$ units, $27\sqrt{3}$ $units^2$ **12.** $32\sqrt{2}$ units, 128 $units^2$ **13.** 2:1, 4:1 **14.** 8:3, 64:9 **15.** 3.98 m **16.** $6\pi \approx 18.85$ ft **17.** 2.62 cm **18.** 6.98 cm **19.** $16\pi \approx 50.27$ $in.^2$ **20.** $30.25\pi \approx 95.03$ m^2 **21.** 16.36 m^2 **22.** 24.74 $in.^2$ **23.** about 40% **24.** about 20% **25.** about 70%

Test B

1. 115 **2.** 360° **3.** 108°, 72° **4.** 135°, 45° **5.** 140°, 40° **6.** 157.5°, 22.5° **7.** 5 **8.** 16 **9.** 12 **10.** 36 **11.** 41.57 units, 83.14 $units^2$ **12.** 87.2 units, 523.1 $units^2$ **13.** 5:2, 25:4 **14.** 5 : 4, 25 : 16 **15.** $\approx$ 21.99 ft **16.** $\approx$ 47.75 cm **17.** 8.51 m **18.** 5.24 in. **19.** 70.88 $in.^2$ **20.** 24.3 $in.^2$ **21.** about 50% **22.** about 37.5% **23.** about 87.5% **24.** about 50% **25.** about 19.4%

Test C

1. 115 **2.** 360° **3.** 60°, 120° **4.** 120°, 60° **5.** 147.3°, 32.7° **6.** 162.9°, 17.1° **7.** 4 **8.** 8 **9.** 12 **10.** 36 **11.** 50 units, 172.05 $units^2$ **12.** 60 units, 259.81 $units^2$ **13.** 3:2, 9:4 **14.** 2:1, 4:1 **15.** $\approx$ 28.27 units **16.** 7.96 m **17.** 3.51 ft **18.** 7.33 in. **19.** about 124.69 m^2 **20.** about 235.06 ft^2 **21.** 18.47 m^2 **22.** 45.55 m^2 **23.** about 37.5% **24.** about 50% **24.** about 75% **26.** about 75% **27.** about 36.3%

Review and Assessment *continued*

SAT/ACT Chapter Test

1. D **2.** A **3.** C **4.** C **5.** B **6.** E **7.** B **8.** A **9.** D **10.** A

Alternative Assessment

1. Complete answers should include:
a. $m\angle JTB = 135°$, $C \approx 62.8$ inches, Length of $\widehat{JB} \approx 23.6$ inches, $A \approx 314$ square inches, Area of sector $JTB = 117.75$ square inches
b. No. When the radius is doubled to 20, the area of sector JTB is 471 square inches. This is four times the original area. **2. a.** 1080° **b.** 135° **c.** 45° **d.** 30.4 inches **e.** 4.62 inches **f.** 70.2 square inches **3. a.** 45.6 inches **b.** about 10.6% **4.** Answers may vary.

Project: Gift Boxes

1. *Sample answer:* Each large circle has radius 6 cm. Each minor arc has measure 90°. Draw a circle with center O and radius 6 cm. Draw diameter $\overline{AE}$. Use the construction on p. 130 to construct diameter $\overline{BD}$ perpendicular to $\overline{AE}$. The endpoints of the diameters (A, B, E, and D) divide the circle into four congruent 90° arcs. Draw an arc with center B and radius 6 cm. Then draw another arc with center E and radius 6 cm. The point where the arcs intersect is the center of the second circle. Construct the second circle. To construct the other three arcs in the second circle, use a procedure similar to the one used for the first circle. **2–5.** Check work.

6–7. The circles overlap by a region that consists of twice the difference of a sector minus a 45°-45°-90° triangle.

	Radius	***Area***
1st box	6 cm	$54\pi + 36 \approx 205.65$
2nd box	9 cm	$\frac{243\pi}{2} + 81 \approx 462.70$
3rd box	12 cm	$216\pi + 144 \approx 822.58$

	$\frac{\textit{radius}}{\textit{radius of 1st box}}$	$\frac{\textit{area}}{\textit{area of 1st box}}$
1st box	–	–
2nd box	$\frac{9}{6} = \frac{3}{2}$	about 2.25, or $\left(\frac{3}{2}\right)^2$
3rd box	$\frac{12}{6} = 2$	about 4, or 2^2

Yes; the ratios of the areas are the squares of the ratios of the radii, as expected because all the patterns are similar.

Cumulative Review

1. 10.8 units, (1, 0)

2.

$7x - 5 - 4x = 4x - 14 - 4x$	Subtraction prop. of equality
$3x - 5 = -14$	Simplify.
$3x - 5 + 5 = -14 + 5$	Additon prop. of equality
$3x = -9$	Simplify.
$\frac{3x}{3} = \frac{-9}{3}$	Division property of equality
$x = -3$	Simplify

3. 15

4.

Statements	Reasons
1. $\overline{BC} \parallel \overline{DA}$	**1.** Given
2. $\angle CBD \cong \angle ABD$	**2.** Alternate Int. ∠s Thm.
3. $\overline{BC} \cong \overline{DA}$	**3.** Given
4. $\overline{BD} \cong \overline{BD}$	**4.** Reflexive prop. of Congruence
5. $\triangle BAD \cong \triangle DCB$	**5.** SAS Congruence Postulate

5. 12 **6.** 122 **7.** (2, −4) **8.** (2, 6) **9.** $x = 46, y = 9$ **10.** $x = 37, y = 10$ **11.** $m\angle A = 68°, m\angle B = 22°$ **12.** $m\angle A = 51°, m\angle B = 39°$ **13.** 110 **14.** 40 **15.** 33.94 units, 72 units^2 **16.** 62.35 units, 187.06 units^2 **17.** 48 units, 166.28 units^2 **18.** 3:2, 9:4 **19.** 4:3, 16:9 **20.** 21.50 units^2 **21.** 34.68 units^2 **22.** 30.96 units^2